DOE-HDBK-1014/1-92
JUNE 1992

DOE FUNDAMENTALS HANDBOOK

MATHEMATICS

Volume 1 of 2

U.S. Department of Energy **FSC-6910**

Washington, D.C. 20585

This document has been reproduced directly from the best available copy.

Available to DOE and DOE contractors from the Office of Scientific and Technical Information. P. O. Box 62, Oak Ridge, TN 37831; (615) 576-8401.

Available to the public from the National Technical Information Service, U.S. Department of Commerce, 5285 Port Royal Rd., Springfield, VA 22161.

Order No. DE92019794

ABSTRACT

The *Mathematics Fundamentals Handbook* was developed to assist nuclear facility operating contractors provide operators, maintenance personnel, and the technical staff with the necessary fundamentals training to ensure a basic understanding of mathematics and its application to facility operation. The handbook includes a review of introductory mathematics and the concepts and functional use of algebra, geometry, trigonometry, and calculus. Word problems, equations, calculations, and practical exercises that require the use of each of the mathematical concepts are also presented. This information will provide personnel with a foundation for understanding and performing basic mathematical calculations that are associated with various DOE nuclear facility operations.

Key Words: Training Material, Mathematics, Algebra, Geometry, Trigonometry, Calculus

www.KnowledgePublications.com

FOREWORD

The *Department of Energy (DOE) Fundamentals Handbooks* consist of ten academic subjects, which include Mathematics; Classical Physics; Thermodynamics, Heat Transfer, and Fluid Flow; Instrumentation and Control; Electrical Science; Material Science; Mechanical Science; Chemistry; Engineering Symbology, Prints, and Drawings; and Nuclear Physics and Reactor Theory. The handbooks are provided as an aid to DOE nuclear facility contractors.

These handbooks were first published as Reactor Operator Fundamentals Manuals in 1985 for use by DOE category A reactors. The subject areas, subject matter content, and level of detail of the Reactor Operator Fundamentals Manuals were determined from several sources. DOE Category A reactor training managers determined which materials should be included, and served as a primary reference in the initial development phase. Training guidelines from the commercial nuclear power industry, results of job and task analyses, and independent input from contractors and operations-oriented personnel were all considered and included to some degree in developing the text material and learning objectives.

The *DOE Fundamentals Handbooks* represent the needs of various DOE nuclear facilities' fundamental training requirements. To increase their applicability to nonreactor nuclear facilities, the Reactor Operator Fundamentals Manual learning objectives were distributed to the Nuclear Facility Training Coordination Program Steering Committee for review and comment. To update their reactor-specific content, DOE Category A reactor training managers also reviewed and commented on the content. On the basis of feedback from these sources, information that applied to two or more DOE nuclear facilities was considered generic and was included. The final draft of each of the handbooks was then reviewed by these two groups. This approach has resulted in revised modular handbooks that contain sufficient detail such that each facility may adjust the content to fit their specific needs.

Each handbook contains an abstract, a foreword, an overview, learning objectives, and text material, and is divided into modules so that content and order may be modified by individual DOE contractors to suit their specific training needs. Each subject area is supported by a separate examination bank with an answer key.

The *DOE Fundamentals Handbooks* have been prepared for the Assistant Secretary for Nuclear Energy, Office of Nuclear Safety Policy and Standards, by the DOE Training Coordination Program. This program is managed by EG&G Idaho, Inc.

www.KnowledgePublications.com

OVERVIEW

The *Department of Energy Fundamentals Handbook* entitled *Mathematics* was prepared as an information resource for personnel who are responsible for the operation of the Department's nuclear facilities. A basic understanding of mathematics is necessary for DOE nuclear facility operators, maintenance personnel, and the technical staff to safely operate and maintain the facility and facility support systems. The information in the handbook is presented to provide a foundation for applying engineering concepts to the job. This knowledge will help personnel more fully understand the impact that their actions may have on the safe and reliable operation of facility components and systems.

The *Mathematics* handbook consists of five modules that are contained in two volumes. The following is a brief description of the information presented in each module of the handbook.

Volume 1 of 2

Module 1 - Review of Introductory Mathematics

This module describes the concepts of addition, subtraction, multiplication, and division involving whole numbers, decimals, fractions, exponents, and radicals. A review of basic calculator operation is included.

Module 2 - Algebra

This module describes the concepts of algebra including quadratic equations and word problems.

Volume 2 of 2

Module 3 - Geometry

This module describes the basic geometric figures of triangles, quadrilaterals, and circles; and the calculation of area and volume.

Module 4 - Trigonometry

This module describes the trigonometric functions of sine, cosine, tangent, cotangent, secant, and cosecant. The use of the pythagorean theorem is also discussed.

Module 5 - Higher Concepts of Mathematics

> This module describes logarithmic functions, statistics, complex numbers, imaginary numbers, matrices, and integral and derivative calculus.

The information contained in this handbook is by no means all encompassing. An attempt to present the entire subject of mathematics would be impractical. However, the *Mathematics* handbook does present enough information to provide the reader with a fundamental knowledge level sufficient to understand the advanced theoretical concepts presented in other subject areas, and to better understand basic system and equipment operations.

Department of Energy
Fundamentals Handbook

MATHEMATICS
Module 1
Review of Introductory Mathematics

www.KnowledgePublications.com

TABLE OF CONTENTS

TABLE OF CONTENTS (Cont.)

TABLE OF CONTENTS (Cont.)

LIST OF FIGURES

LIST OF TABLES

NONE

REFERENCES

- Dolciani, Mary P., et al., Algebra Structure and Method Book 1, Atlanta: Houghton-Mifflin, 1979.
- Naval Education and Training Command, Mathematics, Vol:1, NAVEDTRA 10069-D1, Washington, D.C.: Naval Education and Training Program Development Center, 1985.
- Olivio, C. Thomas and Olivio, Thomas P., Basic Mathematics Simplified, Albany, NY: Delmar, 1977.
- Science and Fundamental Engineering, Windsor, CT: Combustion Engineering, Inc., 1985.
- Academic Program For Nuclear Power Plant Personnel, Volume 1, Columbia, MD: General Physics Corporation, Library of Congress Card #A 326517, 1982.

TERMINAL OBJECTIVE

1.0 Given a basic mathematical problem, **SOLVE** for the answer with or without the aid of a calculator.

ENABLING OBJECTIVES

1.1 **IDENTIFY** the following basic mathematical symbols and definitions.

a.	$=$	equals
b.	$\neq$	is not equal to
c.	$\equiv$	is defined as
d.	$\pm$	plus or minus
e.	$\sqrt[n]{a}$	nth root of a
f.	$\lvert a \rvert$	absolute value of a
g.	$\sum_{i=1}^{N} x_i$	sum of N values
h.	$\angle$	angle
i.	%	percent
j.	x, $\cdot$, *	multiplied by
k.	$\div$, /	divided by
l.	$\geq$	greater than or equal to
m.	$\leq$	less than or equal to
n.	><, <>	is not equal to (computer)
o.	∞	infinity
p.	$\propto$	is proportional to
q.	$\approx$	approximately equal to
r.	$\perp$	perpendicular to
s.	$\parallel$	parallel to

1.2 **APPLY** one of the arithmetic operations of addition, subtraction, multiplication, and division using whole numbers.

1.3 Given a set of numbers, **CALCULATE** the average value.

1.4 **APPLY** one of the arithmetic operations of addition, subtraction, multiplication, and division using fractions.

1.5 **APPLY** one of the arithmetic operations of addition, subtraction, multiplication, and division of fractions by conversion to decimal form using a calculator.

ENABLING OBJECTIVES (Cont.)

1.6 **APPLY** one of the arithmetic operations of addition, subtraction, multiplication, and division using decimals.

1.7 **APPLY** one of the arithmetic operations of addition, subtraction, multiplication, and division using signed numbers.

1.8 **DETERMINE** the number of significant digits in a given number.

1.9 Given a formula, **CALCULATE** the answer with the appropriate number of significant digits.

1.10 **CONVERT** between percents, decimals, and fractions.

1.11 **CALCULATE** the percent differential.

1.12 **APPLY** one of the arithmetic operations of addition, subtraction, multiplication, and division using exponential numbers.

1.13 Given the data, **CONVERT** integers into scientific notation and scientific notation into integers.

1.14 **APPLY** one of the arithmetic operations of addition, subtraction, multiplication, and division to numbers using scientific notation.

1.15 **CALCULATE** the numerical value of numbers in radical form.

TERMINOLOGY

This chapter reviews the terminology and associated symbols used in mathematics.

EO 1.1 IDENTIFY the following basic mathematical symbols and definitions.

a.	$=$	**equals**	**k.**	$\div$, /	**divided by**
b.	$\neq$	**is not equal to**	**l.**	$\geq$	**greater than or equal to**
c.	$\equiv$	**is defined as**	**m.**	$\leq$	**less than or equal to**
d.	$\pm$	**plus or minus**	**n.**	$><$,	**is not equal to (computer)**
e.	$\sqrt[n]{a}$	**nth root of a**		$<>$	
f.	$\lvert a \rvert$	**absolute value of a**	**o.**	∞	**infinity**
g.	$\sum_{i=1}^{N} x_i$	**sum of N values**	**p.**	$\propto$	**is proportional to**
h.	$\angle$	**angle**	**q.**	$\approx$	**approximately equal to**
i.	**%**	**percent**	**r.**	$\perp$	**perpendicular to**
j.	x, $\cdot$, *	**multiplied by**	**s.**	$\parallel$	**parallel to**

In order to understand and communicate in mathematical terms and to lay the foundation for the concepts and principles presented in this material, certain terms and expressions must be defined. This section covers basic definitions used in mathematics. Once understood, such knowledge should provide the foundation from which the principles of mathematics can be presented. By no means are the terms here all inclusive; they are representative of those found within the nuclear field.

Equals

An expression indicating values which are identical in mathematical value or logical denotation. It is given the symbol =.

Is Not Equal to

An expression indicating values which are not identical in mathematical value or logical denotation. It is given the symbol $\neq$ or $><$, $><$ (computer).

Is defined as

A mathematical expression for defining a symbol or variable in mathematics. It is usually given the symbol ≡ .

Plus or Minus

While plus (+) and minus (-) are used individually to indicate addition and subtraction, this form is used to denote a control band, or tolerance band, or error band, such as 100 $\pm$ 5 psig. It is given the symbol $\pm$.

n^{th} root

For any integer (n greater than one), the nth root ($\sqrt[n]{a}$) of a is defined as follows: $\sqrt[n]{a} = b$ if, and only if, $b^n = a$. The number n, in $\sqrt[n]{a}$, is called the index of the root. The nth root of a number (a) is a number (b) which has the property that the product of n values of b is a. For example, the third (or cube) root of 8 is 2, because 2x2x2 equals 8.

Absolute Value of a

This expression represents the magnitude of a variable without regard to its sign. It signifies the distance from zero on a number line. That is, the absolute value of -6 is 6 because -6 is 6 units from zero. Likewise, the absolute value of +6 is 6 because it, too, is 6 units from zero. It is given the symbol $|A|$ where A is any number or variable.

Sum of N values

$\sum_{i=1}^{N} x_i$ indicates the sum of numbered (indexed) values. For example, if the x_i are grades for the individual students in a class, the sum of the x_i (grades) for the students in the class of N students, divided by N, gives the average grade.

Angle

An angle is a set of points consisting of two rays with a common midpoint. It is given the symbol $\angle$.

Percent

An expression used to indicate a fraction of the whole, such as 50% of 90 is 45. It is given the symbol %.

Multiplied by

A mathematical operation that, at its simplest, is an abbreviated process of adding an integer to itself a specified number of times. It is given the symbols x, · , or * (computer).

Divided by
A mathematical process that subjects a number to the operation of finding out how many times it contains another number. It is given the symbol ÷ or /.

Greater than or equal to
It is given the symbol $\geq$, and denotes one quantity is equal to or larger than another.

Less than or equal to
It is given the symbol $\leq$, and denotes one quantity is equal to or smaller than another.

Infinity
A mathematical expression meaning very large in magnitude or distance. It is so large that it cannot be measured. It is given the symbol ∞ .

Is Proportional to
The statement that a is proportionl to b ($a \alpha b$) means that a = (some constant) x b. For example, the dollars you earn in a week (straight rate) are proportional to the hours you work, with the constant being the dollars per hour you earn.

Approximately Equal to
An expression indicating a value which is not exact, but rather close to the value. It is given the symbol $\approx$.

Perpendicular to
This expression means that two objects are at right angles (form a 90-degree angle) to each other. It is given the symbol $\perp$.

Parallel to
Two lines extending in the same direction which are everywhere equidistant and not meeting. It is given the symbol $\parallel$.

Summary

The important information from this chapter is summarized below.

Terminology Summary

- This chapter reviewed the terminology needed in the application and study of mathematics.

CALCULATOR OPERATIONS

This chapter gives the student a chance to reacquaint himself with basic calculator operations.

The teaching of the "mechanics of mathematics" (division, multiplication, logarithms, etc.) in recent years has focused more on the skills of using a calculator than on the pure principles of the basic subject material. With the decreased cost of hand calculators, virtually every person owns, or has access to, a calculator. A nuclear plant operator would be wise to learn how to use most of the calculators available today. Such knowledge will help the operator make quick decisions when circumstances arise for the need of a "quick calculation" of flow rate or some other parameter.

Many calculators are available on the market today, and each one is a little different. For the purpose of this module, a scientific calculator will be needed. The Texas Instruments scientific calculator TI-30 will be used for the examples in this module. Most calculators work on the same principles, but some do not. Some calculators operate on a programming principle like Hewlett-Packard (HP). An HP calculator does not use an equal key. To perform a mathematical operation, the first number is inserted, the ENTER key is pressed, the second number is inserted, and then the mathematical function key is pressed. The result will be displayed. If a different calculator is used, the student will need to refer to the reference manual for his or her calculator.

The following section will review the general use function keys on a TI-30 calculator. In each following chapter of this module, the applicable calculator operations will be addressed.

Appendix A of this module gives a representation of a TI-30 keyboard to assist the student.

Keys

Clear entry/Clear key

[ON/C] Pressing this key once will clear the last operation and the display. Pressing this key twice will clear all operations except the memory.

Note: To clear the memory, press clear then STO.

Note: Many brands break this function into two separate keys, usually labeled "clear" and "all clear," where the "clear" key clears the last entry and the "all clear" key clears the display and all pending operations.

Memory Key

STO

The TI-30 has only one memory. Pressing the STO key enters the displayed number into memory. Any number already in memory will be overwritten.

Note: Calculators with more than one memory will require a number to be entered with the STO key. For example, STO 01 means store the displayed number in memory 01; STO 20 means store the number in memory 20.

Memory Recall Key

RCL

Pressing the RCL key will retrieve the number in memory and display it. Note that the number is also still in memory. This allows the number to be used again. Pressing the RCL will also overwrite any number previously displayed.

Note: Calculators with more than one memory will require a number to be entered with the RCL key. RCL 01 means recall the number stored in the 01 memory. RCL 20 means recall the number stored in memory 20.

Constant Key

K

Certain calculations often contain repetitive operations and numbers. The K, constant, is a time-saving function that allows a single key stroke to perform a single operation and number on the displayed number.

For example, if 20 numbers are to be multiplied by -17.35, the K key can be used. Enter -17.35, then press the times key, then the K key; this "teaches" the calculator the required operation. From this point on when entering a number and pressing the K key, the calculator will automatically multiply the displayed number by -17.35, saving you six key strokes.

Summation Key

SUM

If a long list of numbers is to be added, the summation key will save time if used. Pressing the summation key adds the displayed number to the number in memory. The final sum is then retrieved from memory.

Memory Exchange Key

EXC/$x!$

The EXC, memory exchange key, swaps the displayed number with the number in memory.

Reciprocal Key

$1/x$

When pressed, it divides the displayed number into one.

FOUR BASIC ARITHMETIC OPERATIONS

This chapter reviews the basic mathematical operations of addition, subtraction, multiplication, and division of whole numbers.

EO 1.2 APPLY one of the arithmetic operations of addition, subtraction, multiplication, and division using whole numbers.

Calculator Usage, Special Keys

This chapter requires the use of the +, -, x, ÷, and = keys. When using a TI-30 calculator, the number and operation keys are entered as they are written. For example, the addition of 3 plus 4 is entered as follows:

3 key, + key, 4 key, = key, the answer, 7, is displayed

Parentheses

( )

The parentheses keys allow a complicated equation to be entered as written. This saves the time and effort of rewriting the equation so that multiplication/division is performed first and addition/subtraction is performed second, allowing the problem to be worked from left to right in one pass.

The Decimal Numbering System

The decimal numbering system uses ten symbols called digits, each digit representing a number. These symbols are 0, 1, 2, 3, 4, 5, 6, 7, 8 and 9. The symbols are known as the numbers zero, one, two, three, etc. By using combinations of 10 symbols, an infinite amount of numbers can be created. For example, we can group 5 and 7 together for the number 57 or 2 and 3 together for the number 23. The place values of the digits are multiples of ten and given place titles as follows:

Place Position: 7 6 5 4 3 2 1

Place Position	Place Title	Place Title	
1	Units	1	
2	Tens	10	
3	Hundreds	100	(10x10)
4	Thousands	1000	(10x10x10)
5	Ten Thousands	10,000	(10x10x10x10)
6	Hundred Thousands	100,000	(10x10x10x10x10)
7	Millions	1,000,000	(10x10x10x10x10x10)

Numbers in the decimal system may be classified as integers or fractions. An integer is a whole number such as 1, 2, 3, . . . 10, 11, . . . A fraction is a part of a whole number, and it is expressed as a ratio of integers, such as 1/2, 1/4, or 2/3.

An even integer is an integer which can be exactly divided by 2, such as 4, 16, and 30. All other integers are called odd, such as 3, 7, and 15. A number can be determined to be either odd or even by noting the digit in the units place position. If this digit is even, then the number is even; if it is odd, then the number is odd. Numbers which end in 0, 2, 4, 6, 8 are even, and numbers ending in 1, 3, 5, 7, 9 are odd. Zero (0) is even.

Examples:

Determine whether the following numbers are odd or even: 364, 1068, and 257.

Solution:

1. 364 is even because the right-most digit, 4, is an even number.
2. 1068 is even because the right-most digit, 8, is an even number.
3. 257 is odd because the right-most digit, 7, is an odd number.

Adding Whole Numbers

When numbers are added, the result is called the sum. The numbers added are called addends. Addition is indicated by the plus sign (+). To further explain the concept of addition, we will use a number line to graphically represent the addition of two numbers.

Example: Add the whole numbers 2 and 3.

Solution: Using a line divided into equal segments we can graphically show this addition.

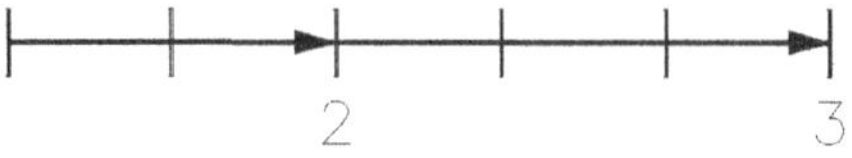

Starting at zero, we first move two places to the right on the number line to represent the number 2. We then move an additional 3 places to the right to represent the addition of the number 3. The result corresponds to the position 5 on the number line. Using this very basic approach we can see that 2 + 3 = 5. Two rules govern the addition of whole numbers.

The commutative law for addition states that two numbers may be added in either order and the result is the same sum. In equation form we have:

$$a + b = b + a \tag{1-1}$$

For example, 5 + 3 = 8 OR 3 + 5 = 8. Numbers can be added in any order and achieve the same sum.

The associative law for addition states that addends may be associated or combined in any order and will result in the same sum. In equation form we have:

$$(a + b) + c = a + (b + c) \tag{1-2}$$

For example, the numbers 3, 5, and 7 can be grouped in any order and added to achieve the same sum:

$$(3 + 5) + 7 = 15 \quad \text{OR} \quad 3 + (5 + 7) = 15$$

The sum of both operations is 15, but it is not reached the same way. The first equation, (3 + 5) + 7 = 15, is actually done in the order (3 + 5) = 8. The 8 is replaced in the formula, which is now 8 + 7 = 15.

The second equation is done in the order (5 + 7) = 12, then 3 + 12 = 15. Addition can be done in any order, and the sum will be the same.

When several numbers are added together, it is easier to arrange the numbers in columns with the place positions lined up above each other. First, the units column is added. After the units column is added, the number of tens is carried over and added to the numbers in the tens column. Any hundreds number is then added to the hundreds column and so on.

Example:

Add 345, 25, 1458, and 6.

Solution:

$$\begin{array}{r} 345 \\ 25 \\ 1458 \\ \underline{+\ 6} \\ 1834 \end{array}$$

When adding the units column, 5 + 5 + 8 + 6 = 24. A 4 is placed under the units column, and a 2 is added to the tens column.

Then, 2 + 4 + 2 + 5 = 13. A 3 is placed under the tens column and a 1 is carried over to the hundreds column. The hundreds column is added as follows: 1 + 3 + 4 = 8.

An 8 is placed under the hundreds column with nothing to carry over to the thousands column, so the thousands column is 1. The 1 is placed under the thousands column, and the sum is 1834. To verify the sum, the numbers should be added in reverse order. In the above example, the numbers should be added from the bottom to the top.

Subtracting Whole Numbers

When numbers are subtracted, the result is called the remainder or difference. The number subtracted is called the subtrahend; the number from which the subtrahend is subtracted is called the minuend. Subtraction is indicated by the minus sign (-).

86	Minuend
-34	-Subtrahend
52	Remainder or Difference

Unlike addition, the subtraction process is neither associative nor commutative. The commutative law for addition permitted reversing the order of the addends without changing the sum. In subtraction, the subtrahend and minuend cannot be reversed.

$$a - b \neq b - a \tag{1-3}$$

Thus, the difference of 5 - 3 is not the same as 3 - 5. The associative law for addition permitted combining addends in any order. In subtraction, this is not allowed.

$$(a-b)-c \neq a-(b-c)$$

Example:

$$(10-5)-1 \neq 10-(5-1)$$
$$4 \neq 6$$

When subtracting two numbers, the subtrahend is placed under the minuend with the digits arranged in columns placing the units place under the units place, the tens column next, and so on.

Example:

Subtract 32 from 54.

Solution:

$$\begin{array}{r} 54 \\ \underline{-32} \\ 22 \end{array}$$

Whenever the digit in the subtrahend is larger than the digit in the minuend in the same column, one place value is borrowed from the next digit to the left in the minuend. Refer to the following example.

Example:

Subtract 78 from 136.

Solution:

$$\begin{array}{r} {\scriptstyle 2} \\ 1\not{3}6 \\ \underline{-\ 78} \\ 58 \end{array}$$

When subtracting the units column, 6 - 8, a 10 is borrowed from the tens column. This now makes subtracting the units column 16 - 8. An 8 is placed under the units column. Next the tens column is subtracted.

A 10 was borrowed from the tens column and now 7 is subtracted from 12, not 13. This yields: 12 - 7 = 5. The 5 is placed under the tens column and the difference is 58.

This can be done for any subtraction formula. When the digit in the subtrahend column is larger than the digit in the minuend in the same column, a number from the next higher place position column is "borrowed." This reduces the higher position column by one.

Subtraction can be verified by adding the difference to the subtrahend, which should result in the minuend.

Multiplying Whole Numbers

Multiplication is the process of counting a number two or more times. It can be considered a shortened form of addition. Thus, to add the number 4 three times, 4 + 4 + 4, we can use multiplication terms, that is, 4 multiplied by 3. When numbers are multiplied, the result is called the product. The numbers multiplied are called factors. One factor is called the multiplicand; the other is called the multiplier. Multiplication is indicated by the times or multiplication sign (x), by a raised dot (·), or by an asterick (*).

9	Multiplicand
x 4	x Multiplier
36	Product

In multiplying several numbers, the same product is obtained even if the numbers are multiplied in a different order or even if some of the numbers are multiplied together before the final multiplication is made. These properties are called the commutative and associative laws for multiplication.

The commutative law for multiplication states that numbers can be multiplied in any order, and the result is the same product. In equation form:

$$a \text{ x } b = b \text{ x } a \tag{1-4}$$

Thus, the product of 8 x 3 is the same as 3 x 8.

The associative law for multiplication states that factors can be associated in any order, and the result is the same product. In equation form:

$$a \text{ x } (b \text{ x } c) = (a \text{ x } b) \text{ x } c \tag{1-5}$$

Thus, the numbers 2, 3, and 5 can be multiplied by first multiplying 2 x 3 to equal 6 and then multiplying 6 x 5 to equal 30. The equation may also be solved by first multiplying 3 x 5 to equal 15, and then multiplying 15 x 2 to equal 30. In either case, the product is 30.

In multiplying two numbers, one number is placed under the other with the digits arranged in columns placing units under the units place, tens under the tens place, and so on. Usually, the larger number is considered the multiplicand and the smaller number is considered the multiplier. The digit in the units place of the multiplier is multiplied first, the digit in the tens place of the multiplier next, and so on.

Example 1:

Multiply 432 by 8.

Solution:

$$\begin{array}{r} 432 \\ \times\ 8 \\ \hline 3{,}456 \end{array}$$

In multiplying the multiplier in the units column to the multiplicand, 8 x 2 = 16. A 6 is placed under the units column, and 1 ten is carried. Then, 8 x 3 = 24, plus the 1 carried over equals 25.

A 5 is placed under the tens column, and 2 hundreds are carried over. Next, 8 x 4 = 32, plus 2 carried over, equals 34. A 4 is placed under the hundreds column and a 3 under the thousands column.

Example 2:

What is the product of 176 x 59?

Solution:

$$\begin{array}{rl} 176 & \\ \text{x } 59 & \\ \hline 1584 & \text{Multiplication by 9} \\ 880 & \text{Multiplication by 50} \\ \hline 10384 & \end{array}$$

Start by multiplying the digit in the units place of the multiplier, 9 x 6 = 54. A 4 is placed under the units column, and 5 tens are carried over.

Next, 9 x 7 = 63, plus the 5 carried over, equals 68. An 8 is placed under the tens column, and 6 hundreds are carried over. Then, 9 x 1 = 9, plus 6 carried over, equals 15. A 5 is placed under the hundreds column and a 1 under the thousands column.

The digit in the tens place of the multiplier is multiplied now: 5 x 6 = 30. Since the 5 in 59 is in the tens column, the zero is placed under the tens column, and 3 tens are carried over. Next, 5 x 7 = 35, plus the 3 carried over, equals 38. An 8 is placed under the hundreds column, and 3 hundreds are carried over.

Then, 5 x 1 = 5, plus 3 carried over, equals 8. An 8 is placed under the thousands column. The results of 176 multiplied by 9 and 50 are then added to give the final product.

Dividing Whole Numbers

Division is the process of determining how many times one number is contained in another number. When numbers are divided, the result is the quotient and a remainder. The remainder is what remains after division. The number divided by another number is called the dividend; the number divided into the dividend is called the divisor. Division is indicated by any of the following:

a division sign (÷)

a division sign ($\overline{)\quad}$)

a horizontal line with the dividend above the line and the divisor below the line $\left(\frac{\#}{\#}\right)$

a slanting line a/b meaning a divided by b

Thus, the relationship between the dividend, divisor, and quotient is as shown below:

37	Dividend
÷ 4	Divisor
9	Quotient
1	Remainder

Quotient → 9 1 ← Remainder
$4\overline{)37}$
Divisor ↑ ↑Dividend

Unlike multiplication, the division process is neither associative nor commutative. The commutative law for multiplication permitted reversing the order of the factors without changing the product. <u>In division the dividend and divisor cannot be reversed</u>.

Using the equation form:

$$a \div b \neq b \div a \tag{1-6}$$

For example, the quotient of 18 ÷ 6 is not the same as the quotient of 6 ÷ 18. 18 divided by 6 equals 3; 6 divided by 18 equals 0.33.

The associative law for multiplication permitted multiplication of factors in any order. In division, this is not allowed.

$$(a \div b) \div c \neq a \div (b \div c)$$

Example: $(8 \div 4) \div 2 \neq 8 \div (4 \div 2)$

$$1 \neq 4$$

When dividing two numbers, the divisor and dividend are lined up horizontally with the divisor to the left of the dividend. Division starts from the left of the dividend and the quotient is written on a line above the dividend.

Example 1:

Divide 347 by 5.

Solution:

```
     69
  5) 347
     30
     ---
      47
      45
      --
       2   Remainder
```

Starting from the left of the dividend, the divisor is divided into the first digit or set of digits it divides into. In this case, 5 is divided into 34; the result is 6, which is placed above the 4.

This result (6) is then multiplied by the divisor, and the product is subtracted from the set of digits in the dividend first selected. 6 x 5 equals 30; 30 subtracted from 34 equals 4.

The next digit to the right in the dividend is then brought down, and the divisor is divided into this number. In this case, the 7 is brought down, and 5 is divided into 47; the result is 9, which is placed above the 7.

Again, this result is multiplied by the divisor, and the product is subtracted from the last number used for division. 9 x 5 equals 45; 45 subtracted from 47 equals 2. This process is repeated until all of the digits in the dividend have been brought down. In this case, there are no more digits in the dividend. The result of the last subtraction is the remainder. The number placed above the dividend is the quotient. In this case, 347 ÷ 5 yields a quotient of 69 with a remainder of 2.

Example 2:

Divide 738 by 83.

Solution:

```
      8
83) 738
    664
    ---
     74     Remainder
```

Example 3:

Divide 6409 by 28.

Solution:

```
      228
28) 6409
    56
    --
     80
     56
     --
     249
     224
     ---
      25     Remainder
```

Division can be verified by multiplying the quotient by the divisor and adding the remainder. The result should be the dividend. Using Example 3, multiply 228 by 28 to check the quotient.

```
  228
x  28
-----
 1824
 456
-----
 6384  →  Product
+  25  →  Remainder of 25
-----
 6409
```

Hierarchy of Mathematical Operations

Mathematical operations such as addition, subtraction, multiplication, and division are usually performed in a certain order or sequence. Typically, multiplication and division operations are done prior to addition and subtraction operations. In addition, mathematical operations are also generally performed from left to right using this heirarchy. The use of parentheses is also common to set apart operations that should be performed in a particular sequence.

Example:

Perform the following mathematical operations to solve for the correct answer:

$$(2 + 3) + (2 \times 4) + \left(\frac{6+2}{2} \right) = \underline{\qquad\qquad}$$

Solution:

a. Mathematical operations are typically performed going from left to right within an equation and within sets of parentheses.

b. Perform all math operations within the sets of parentheses first.

$2 + 3 = 5$
$2 \times 4 = 8$
$\frac{6+2}{2} = \frac{8}{2} = 4$ Note that the addition of 6 and 2 was performed prior to dividing by 2.

c. Perform all math operations outside of the parentheses. In this case, add from left to right.

$5 + 8 + 4 = 17$

Example:

Solve the following equation:

$$(4 - 2) + (3 \times 4) - (10 \div 5) - 6 = \underline{\qquad}$$

Solution:

a. Perform math operations inside each set of parentheses.

$4 - 2 = 2$
$3 \times 4 = 12$

$10 \div 5 = 2$

b. Perform addition and subtraction operations from left to right.

c. The final answer is $2 + 12 - 2 - 6 = 6$

There may be cases where several operations will be performed within multiple sets of parentheses. In these cases you must perform all operations within the innermost set of parentheses and work outward. You must continue to observe the hierarchical rules through out the problem. Additional sets of parentheses may be indicated by brackets, [ ].

Example:

Solve the following equation:

$[2 \cdot (3 + 5) - 5 + 2] \times 3 =$ ______

Solution:

a. Perform operations in the innermost set of parentheses.

$3 + 5 = 8$

b. Rewriting the equation:

$[2 \cdot 8 - 5 + 2] \times 3 =$

c. Perform multiplication prior to addition and subtraction within the brackets.

$[16 - 5 + 2] \times 3 =$
$[11 + 2] \times 3 =$
$[13] \times 3 =$

d. Perform multiplication outside the brackets.

$13 \times 3 = 39$

Example:

Solve the following equation:

5 + [2 (3 + 1) - 1] x 2 = _____

Solution:

5 + [2 (4) - 1] x 2 =
5 + [8 - 1] x 2 =
5 + [7] x 2 =
5 + 14 = 19

Example:

Solve the following equation:

[(10 - 4) ÷ 3] + [4 x (5 - 3)] = _____

Solution:

[(6) ÷ 3] + [4 x (2)] =
[2] + [8] =
2 + 8 = 10

Summary

The important information from this chapter is summarized below.

Four Basic Arithmetic Operations Summary

This chapter reviewed using whole numbers to perform the operations of:

- Addition
- Subtraction
- Multiplication
- Division

While this chapter discussed the commutative and associative laws for whole numbers, it should be noted that these laws will also apply to the other types of numbers discussed in later chapters and modules of this course.

AVERAGES

This chapter covers the concept of averages and how to calculate the average of a given set of data.

EO 1.3 Given a set of numbers, CALCULATE the average value.

An average is the sum of a group of numbers or quantities divided by the number of numbers or quantities. Averages are helpful when summarizing or generalizing a condition resulting from different conditions. For example, when analyzing reactor power level, it may be helpful to use the average power for a day, a week, or a month. The average can be used as a generalization of the reactor power for the day, week, or month.

Average calculations involve the following steps:

Step 1: Add the individual numbers or quantities.
Step 2: Count the number of numbers or quantities.
Step 3: Divide the sum in Step 1 by the number in Step 2.

Example 1:

Find the average cost of a car, given the following list of prices.
$10,200; $11,300; $9,900; $12,000; $18,000; $7,600

Solution:

Step 1: 10200 + 11300 + 9900 + 12000 + 18000 + 7600 = 69000
Step 2: Total number of prices is 6
Step 3: Divide 69000 by 6. The result is 11500

Thus, the average price of the six cars is $11,500.

Example 2:

Find the average temperature if the following values were recorded: 600°F, 596°F, 597°F, 603°F

Solution:

Step 1: 600 + 596 + 597 + 603 = 2396
Step 2: The number of items is 4.
Step 3: 2396/4 = 599°F

Average Value

The summation symbol, Σ, introduced in the first chapter, is often used when dealing with the average value, $\bar{x}$.

Using the first example in this chapter, the average value could have been expressed in the following manner:

$$\bar{x}_{car} = \frac{\sum_{i=1}^{N} x_i}{N}$$

where:

$\bar{x}_{car}$ = the average value (cost) of a car

x_i = each of the individual car prices

N = total number of cars

The right side of the above equation can then be rewritten.

$$\bar{x}_{car} = \frac{x_1 + x_2 + x_3 + x_4 + x_5 + x_6}{6}$$

substituting 10,200 for x_1, 11300 for x_2, 9,900 for x_3, etc.

$$\bar{x}_{car} = \frac{10,200 + 11,300 + 9,900 + 12,000 + 18,000 + 7600}{6}$$

$$\bar{x}_{car} = 11,500$$

Example:

If we were to apply the average value equation from above to the second example concerning temperature, how would it be written, and what would be the values for N_1, x_i?

Solution:

$$\bar{x}_{temp} = \frac{\sum_{i=1}^{4} x_i}{4}$$

$x_1 = 600$
$x_2 = 596$
$x_3 = 597$
$x_4 = 603$

$$\bar{x}_{temp} = \frac{x_1 + x_2 + x_3 + x_4}{4}$$

$$= \frac{600 + 596 + 597 + 603}{4}$$

$$= 599$$

Summary

The important information from this chapter is summarized below.

Averages Summary

Calculating the average of a set of numbers requires three steps:

1. Add the individual numbers or quantities.
2. Count the number of numbers or quantities added in previous step.
3. Divide the sum calculated in step 1 by the number in step 2.

FRACTIONS

This chapter covers the basic operations of addition, subtraction, multiplication, and division using fractions.

EO 1.4 APPLY one of the arithmetic operations of addition, subtraction, multiplication, and division using fractions.

A common fraction, such as $\frac{1}{3}$, consists of the numerator 1 and the denominator 3. It is referred to as a rational number describing the division of 1 by 3 (division of the numerator by the denominator).

Proper and Improper Fractions

There are two types of fractions: proper fractions and improper fractions. The value of the numerator and the denominator determines the type of fraction. If the numerator is less than the denominator, the fraction is less than one; this fraction is called a proper fraction. If the numerator is equal to or greater than the denominator, the fraction is called an improper fraction.

Example:

$\frac{3}{8}$ proper fraction

$\frac{8}{3}$ improper fraction

$\frac{3}{3}$ improper fraction

An improper fraction expressed as the sum of an integer and a proper fraction is called a mixed number.

To write an improper fraction as a mixed number, divide the numerator by the denominator, obtaining an integer part (quotient) plus a fractional part whose numerator is the remainder of the division.

Example:

$$\frac{22}{9} = 2 + \frac{4}{9} = 2\frac{4}{9}$$

Here, 9 can be divided into 22 two times, with $\frac{4}{9}$ left over or remaining.

Thus, the improper fraction $\frac{22}{9}$ is equivalent to the mixed number $2\frac{4}{9}$.

Every number may be expressed as a fraction or sum of fractions. A whole number is a fraction whose denominator is 1. Any fraction with the same numerator and denominator is equal to one.

Examples:

$$5 = \frac{5}{1}, \ \frac{10}{1} = 10, \ 1 = \frac{16}{16}, \ \frac{5}{5} = 1$$

Equivalent Fractions

An equivalent fraction is a fraction that is equal to another fraction.

Example:

$$\frac{2}{3} = \frac{4}{6} = \frac{6}{9}$$

A fraction can be changed into an equivalent fraction by multiplying or dividing the numerator and denominator by the same number.

Example:

$\frac{2}{3} \cdot \frac{2}{2} = \frac{4}{6}$ because $\frac{2}{2} = 1$, and 1 x any number = that number

A fraction may be reduced by dividing both the numerator and the denominator of a fraction by the same number.

Example:

$$\frac{6}{8} = \frac{\frac{6}{2}}{\frac{8}{2}} = \frac{3}{4}$$

Addition and Subtraction of Fractions

When two or more fractions have the same denominator, they are said to have a common denominator. The rules for adding fractions with a common denominator will first be explored. Consider the example.

$$\frac{3}{8} + \frac{1}{8} = \underline{\quad\quad}$$

First of all, the fraction $\frac{3}{8}$ means three $\frac{1}{8}$ segments, i.e. $\frac{3}{8} = 3 \text{ x } \frac{1}{8}$. Looking at this as the addition of pie segments:

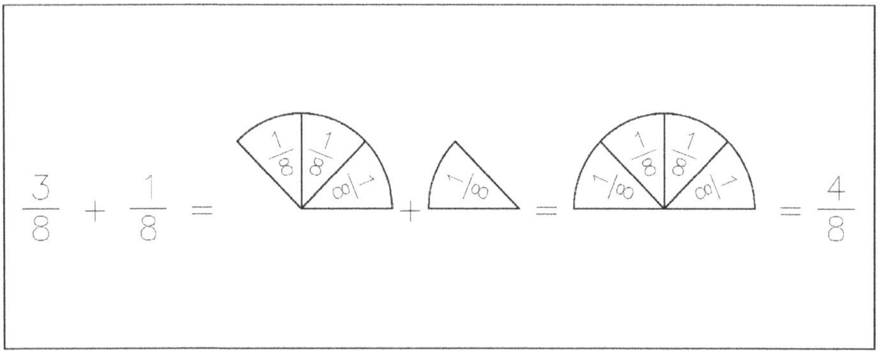

It is obvious that three of these segments $\left(\frac{1}{8} ths\right)$ plus one of these segments $\left(\frac{1}{8} ths\right)$ equal four of these segments $\left(\frac{1}{8} ths\right)$.

This graphic illustration can be done for any addition of fractions with common denominators. The sum of the fractions is obtained by adding the numerators and dividing this sum by the common denominator.

$$\frac{2}{6}+\frac{3}{6}+\frac{1}{6} = 2\times\left(\frac{1}{6}\right)+3\times\left(\frac{1}{6}\right)+1\times\left(\frac{1}{6}\right) = 6\times\frac{1}{6} = 1$$

Also, this general method applies to subtraction, for example,

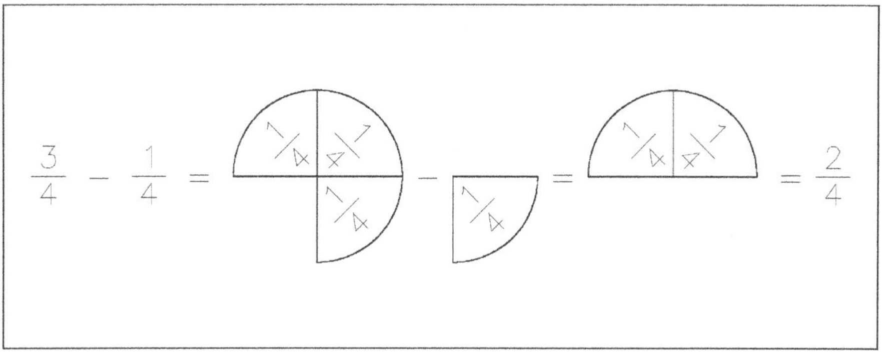

The general method of subtraction of fractions with common denominators is to subtract the numerators and place this difference over the common denominator.

$$\frac{5}{8}-\frac{2}{8} = 5\times\left(\frac{1}{8}\right) - 2\times\left(\frac{1}{8}\right) = (5-2)\times\left(\frac{1}{8}\right) = 3\times\left(\frac{1}{8}\right) = \frac{3}{8}$$

When fractions do not have a common denominator, this method must be modified. For example, consider the problem:

$$\frac{1}{2} + \frac{1}{3} = ?$$

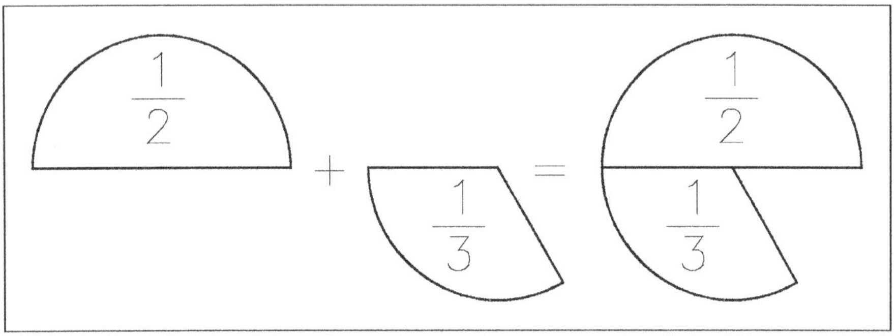

This presents a problem, the same problem one would have if he were asked to add 6 feet to 3 yards. In this case the entities (units) aren't equal, so the 6 feet are first converted to 2 yards and then they are added to 3 yards to give a total of 5 yards.

6 feet + 3 yards = 2 yards + 3 yards = 5 yards

Going back to the fraction addition example, then $\frac{1}{2}$ and $\frac{1}{3}$ must both be expressed in the same segments to be added. Without developing the general method, $\frac{1}{2}$ is $\frac{3}{6}$*ths*. Multiply $\frac{1}{2}$ by $\frac{3}{3}$ or (one) to give the equivalent fraction. Similarly, $\frac{1}{3}$ equals $\frac{2}{6}$.

Then,

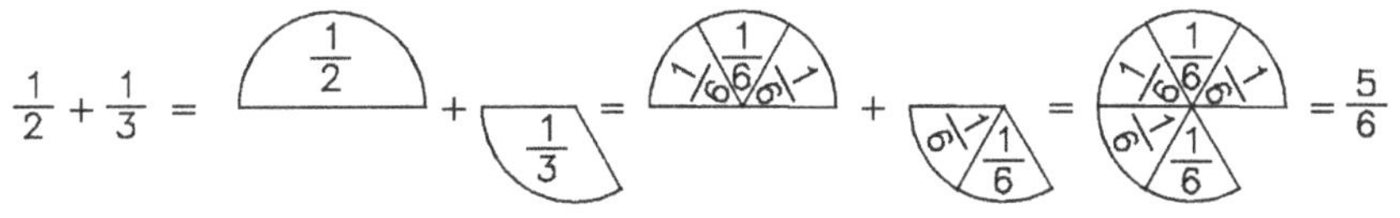

The general method of adding or subtracting fractions which do not have a common denominator is to convert the individual fractions to equivalent fractions with a common denominator. These equally sized segments can then be added or subtracted.

The simplest method to calculate a common denominator is to multiply the denominators. This is obtained if each fraction is multiplied top and bottom by the denominator of the other fraction (and thus by one, giving an equivalent fraction).

$$\frac{1}{3} + \frac{8}{6} =$$

$$\frac{1}{3} \cdot \frac{6}{6} + \frac{8}{6} \cdot \frac{3}{3} =$$

$$\frac{6}{18} + \frac{24}{18} = \frac{30}{18}$$

For more than two fractions, each fraction is multiplied top and bottom by each of the other denominators. This method works for simple or small fractions. If the denominators are large or many fractions are to be added, this method is cumbersome.

Example:

$$\frac{105}{64} + \frac{15}{32} + \frac{1}{6} = \_\_\_\_$$

would require the denominator to be equal to 64 x 32 x 6 = 12,288. This kind of number is very hard to use.

In the earlier example

$\frac{1}{3} + \frac{8}{6} =$ was shown to equal

$$\frac{6}{18} + \frac{24}{18} = \frac{30}{18}.$$

You notice that both 30 and 18 can be divided by 6; if this is done:

$$\frac{30 \div 6}{18 \div 6} = \frac{5}{3}$$

By doing this we arrive at a smaller and more useful number: $\frac{5}{3}$ takes the place of $\frac{30}{18}$.

The sum of two or more fractions reduced to its simplest form contains the smallest possible denominator common to both fractions. This denominator is called the least common denominator (LCD).

Example:

$$\frac{1}{3} + \frac{1}{6} + \frac{1}{8} = \_\_\_\_$$

Using trial and error we can find that 24 is the LCD or smallest number that 3, 6, and 8 will all divide into evenly. Therefore, if each fraction is converted into 24^{ths}, the fractions can be added.

$$\frac{1}{3} \cdot \left(\frac{8}{8}\right) + \frac{1}{6} \cdot \left(\frac{4}{4}\right) + \frac{1}{8} \cdot \left(\frac{3}{3}\right) =$$

$$\frac{8}{24} + \frac{4}{24} + \frac{3}{24} = \frac{15}{24}$$

This is the simplest form the fraction can have. To eliminate the lengthy process of trial and error used in finding the LCD, you can reduce the denominators to their prime numbers.

Least Common Denominator Using Primes

A prime number is a whole number (integer) whose only factors are itself and one. The first prime numbers are:

1, 2, 3, 5, 7, 11, 13, 17, 19, 23, 29, ...

By dividing by primes, you can find that the primes of 105 are:

$\frac{105}{3} = 35 \quad \frac{35}{5} = 7 \quad 7 =$ a prime number, therefore, stop dividing.

The primes of 105 are: 3, 5, 7

A systematic way of finding the prime factors of larger positive integers is illustrated below. The primes are tried in order, as factors, using each as many times as possible before going on to the next. The result in this case is:

$$\begin{aligned} 504 &= (2)(252) \\ &= (2)(2)(126) \\ &= (2)(2)(2)(63) \\ &= (2)(2)(2)(3)(21) \\ &= (2)(2)(2)(3)(3)(7) \end{aligned}$$

To add several fractions with different denominators, follow these steps:

Step 1: Express denominators in prime factors.

Step 2: Determine the least common denominator by using all of the prime numbers from the largest denominator, and then include each prime number from the other denominators so that each denominator can be calculated from the list of primes contained in the LCD.

Step 3: Rewrite using the least common denominator.

Step 4: Add the fractions.

Example 1:

Add $\frac{1}{15}$ and $\frac{7}{10}$

Solution:

Step 1: Find primes of each denominator.

$$15 = 5 \times 3$$
$$10 = 5 \times 2$$

Step 2: In the example, 15 is the largest denominator, so use the 5 and the 3; now look at the second denominator's primes—the five already appears in the list, but the 2 does not, so use the 2.

$$5 \times 3 \times 2 = 30$$

Step 3: Rewrite with least common denominators.

$$\frac{1}{15} = \frac{2}{30}$$
$$\frac{7}{10} = \frac{21}{30}$$

Step 4: Add the new fractions.

$$\frac{2}{30} + \frac{21}{30} = \frac{23}{30}$$

Example 2:

Add $\frac{1}{7} + \frac{2}{3} + \frac{11}{12} + \frac{4}{6}$

Solution:

Step 1: Find primes of each denominator.

7 = 7 (already is a prime number)
3 = 3 (already is a prime number)
$12 = 2 \times 6 = 2 \times 2 \times 3$
$6 = 2 \times 3$

Step 2: 12 is the largest, so start with
$2 \times 2 \times 3$

Comparing this list to the others, the denominators of 3, 12, and 6 can all be calculated from the list, but 7 cannot be, so a 7 must be included in the list.

$$2 \times 2 \times 3 \times 7 = 84$$

Step 3: Rewrite the equation

$$\frac{1}{7} \cdot \frac{12}{12} + \frac{2}{3} \cdot \frac{28}{28} + \frac{11}{12} \cdot \frac{7}{7} + \frac{4}{6} \cdot \frac{14}{14} =$$

Step 4: Add

$$\frac{12}{84} + \frac{56}{84} + \frac{77}{84} + \frac{56}{84} = \frac{201}{84}$$

Addition and Subtraction

Denominators of fractions being added or subtracted must be the same.

The resulting sum or difference is then the sum or difference of the numerators of the fractions being added or subtracted.

Examples:

$$\frac{2}{3} + \frac{1}{3} = \frac{2 + 1}{3} = 1$$

$$\frac{4}{7} + \frac{1}{7} = \frac{4 + 1}{7} = \frac{5}{7}$$

$$\frac{4}{9} - \frac{2}{9} = \frac{4 - 2}{9} = \frac{2}{9}$$

$$\frac{8}{11} - \frac{2}{11} - \frac{5}{11} = \frac{8 - 2 - 5}{11} = \frac{1}{11}$$

Multiplication

The methods of multiplication of fractions differ from addition and subtraction. The operation of multiplication is performed on both the numerator and the denominator.

Step 1: Multiply the numerators.

Step 2: Multiply the denominators.

Step 3: Reduce fraction to lowest terms.

Example:

$$\frac{2}{3} \cdot \frac{1}{4} = \frac{2}{12} = \frac{1}{6}$$

Multiplication of mixed numbers may be accomplished by changing the mixed number to an improper fraction and then multiplying the numerators and denominators.

Example:

$$1\frac{1}{2} \cdot \frac{3}{5} = \frac{3}{2} \cdot \frac{3}{5} = \frac{9}{10}$$

Division

The division of fractions can be performed by two methods. The first method employs the basic concept of multiplying by 1.

Example:

$$\frac{\left(\frac{4}{5}\right)}{\left(\frac{2}{9}\right)} = \underline{\qquad}$$

Solution:

Step 1: Multiply by $\frac{\left(\frac{9}{2}\right)}{\left(\frac{9}{2}\right)}$, which is the same as multiplying by 1.

$$\frac{\left(\frac{4}{5}\right)}{\left(\frac{2}{9}\right)} \cdot \frac{\left(\frac{9}{2}\right)}{\left(\frac{9}{2}\right)} = \_\_\_\_$$

Step 2: Looking at the two division fractions we see that $\frac{2}{9} \cdot \frac{9}{2} = 1$. This leaves us with the following.

$$\frac{\frac{4}{5} \cdot \frac{9}{2}}{1} = \frac{4}{5} \cdot \frac{9}{2}$$

Step 3: Multiply numerators and denominators.

$$\frac{4}{5} \cdot \frac{9}{2} = \frac{36}{10}$$

Example:

$$\frac{\left(\frac{3}{8}\right)}{\left(\frac{6}{7}\right)} = \_\_\_\_$$

Solution:

Step 1: Multiply by $\frac{\frac{7}{6}}{\frac{7}{6}}$.

$$\frac{\frac{3}{8} \cdot \frac{7}{6}}{\frac{6}{7} \cdot \frac{7}{6}} =$$

Step 2: Multiplication of division fractions equals 1.

$$\frac{\frac{3}{8} \cdot \frac{7}{6}}{1} =$$

Step 3: Multiplication of numerators and denominators yields:

$$\frac{3}{8} \cdot \frac{7}{6} = \frac{21}{48}$$

The second method for dividing fractions is really a short cut to the first method. When dividing one fraction by another, first invert the divisor fraction and then multiply.

Example:

$$\frac{\left(\frac{4}{5}\right)}{\left(\frac{2}{9}\right)} =$$

Solution:

Step 1: Invert the divisor fraction $\left(\frac{2}{9}\right)$ to $\left(\frac{9}{2}\right)$.

Step 2: Multiply the dividend fraction, $\left(\frac{4}{5}\right)$, by the inverted fraction $\left(\frac{9}{2}\right)$.

$$\left(\frac{4}{5}\right) \cdot \left(\frac{9}{2}\right) = \frac{36}{10}$$

Step 3: Reduce fraction to lowest terms.

$$\frac{\left(\frac{36}{2}\right)}{\left(\frac{10}{2}\right)} = \frac{18}{5} = 3\frac{3}{5}$$

Division of mixed numbers may be accomplished by changing the mixed number into an improper fraction (*a*/*b*), inverting the divisor, and proceeding as in multiplication.

Invert the divisor fraction and then follow the rule for multiplication.

Example:

$$\frac{1\frac{2}{3}}{\frac{3}{7}} = \frac{5}{3} \cdot \frac{7}{3} = \frac{35}{9} = 3\frac{8}{9}$$

Summary

The important information from this chapter is summarized below.

Fractions Summary

Denominator - bottom number in a fraction

Numerator - top number in a fraction

Proper fraction - numerator is less than denominator

Improper fraction - numerator is greater than or equal to denominator

Mixed number - sum of an integer and a proper fraction

Fractions, like whole numbers can be:

a. Added

b. Subtracted

c. Multiplied

d. Divided

DECIMALS

This chapter covers the processes of addition, subtraction, multiplication, and division of numbers in decimal form.

EO 1.5 **APPLY one of the arithmetic operations of addition, subtraction, multiplication, and division of fractions by conversion to decimal form using a calculator.**

EO 1.6 **APPLY one of the arithmetic operations of addition, subtraction, multiplication, and division using decimals.**

When using numbers, the operator will use whole numbers at times and decimal numbers at other times. A decimal number is a number that is given in decimal form, such as 15.25. The decimal portion is equivalent to a certain "fraction-of-one," thus allowing values between integer numbers to be expressed.

A decimal is a linear array of integers that represents a fraction. Every decimal place indicates a multiple of a power of 10.

Example:

the decimal 0.1 = 1/10, 0.12 = 12/100, and 0.003 = 3/1000

Fraction to Decimal Conversion

In the process of converting a fraction to a decimal, we must perform the operation of division that the fraction represents.

Example:

Convert $\frac{3}{4}$ to a decimal.

Solution:

The fraction $\frac{3}{4}$ represents 3 divided by 4. To put this into decimal form, we first divide 3 by 4. Add a decimal point and zeros to carry out this division.

$$\begin{array}{r} 0.75 \\ 4\overline{)\,3.00} \\ \underline{2\,8} \\ 20 \\ \underline{20} \\ 0 \end{array}$$

Example:

Convert $\frac{1}{3}$ to a decimal.

Solution:

$$\begin{array}{r} 0.33\overline{3} \\ 3\overline{)\,1.000} \\ \underline{9} \\ 10 \\ \underline{9} \\ 10 \\ \underline{9} \\ 1 \end{array}$$

In the above example we see that no matter how many zeros we add, there will always be a remainder of 1. This is called a repeating decimal. A repeating decimal is indicated by a dash over the last number to the right of the decimal point. So, $\frac{1}{3} = 0.33\overline{3}$. The bar is placed over the repeating portion. For a repeating single digit, the bar is placed over only a single digit. For a repeating sequence of digits, the bar is placed over the whole sequence of digits.

Decimal to Fraction Conversion

The process of decimal to fraction conversion involves the use of the fundamental rule of fractions; the fraction should be written in its lowest terms. The following examples demonstrate how to convert decimals to fractions.

Example 1:

Convert 0.65 to a fraction.

Solution:

Step 1: Note the number of place positions to the right of the decimal point. In this example, 0.65 is 65 hundredths, which is two places to the right of the decimal point.

$$\frac{65}{100}$$

Step 2: Although we have now converted the decimal into a fraction, the fraction is not in its lowest terms. To reduce the new fraction into its lowest or simplest terms, both the numerator and the denominator must be broken down into primes.

$$\frac{65}{100} = \frac{5 \cdot 13}{5 \cdot 20} = \frac{5 \cdot 13}{5 \cdot 4 \cdot 5} = \frac{5 \cdot 13}{5 \cdot 2 \cdot 2 \cdot 5}$$

Note that we can cancel one set of 5s, because $\frac{5}{5} = 1$.

This gives

$$\frac{65}{100} = \frac{13}{20}$$

and this is the simplest form of this fraction.

Example 2:

Convert 18.82 to a mixed number.

Solution:

Step 1: 18.82 is 18 and 82 hundredths.

$$18.82 = 18\frac{82}{100}$$

Step 2: Reduce $\frac{82}{100}$ to its simplest form $= \frac{41}{50}$

$$\frac{82}{100} = \frac{2 \cdot 41}{2 \cdot 50} = \frac{2 \cdot 41}{2 \cdot 2 \cdot 25} = \frac{2 \cdot 41}{2 \cdot 2 \cdot 5 \cdot 5} = \frac{41}{2 \cdot 5 \cdot 5} = \frac{41}{50}$$

The answer is $18\frac{41}{50}$.

Example 3:

Convert 1.73 to a fraction.

Solution:

Step 1: $1.73 = 1\frac{73}{100}$

Step 2: 73 = 73 x 1
100 = 2 x 2 x 5 x 5

There are no common factors between 73 and 100, so it cannot be reduced.

$$1\frac{73}{100}$$

Example 4:

Convert 0.333 to a fraction.

Solution:

Step 1: $0.333 = \frac{333}{1000}$

Step 2: There are no common factors between 333 and 1000, so it is already in its simplest form.

Addition and Subtraction of Decimals

When adding or subtracting decimals, each number must be placed to align the decimal points. When necessary, zeros are used as place holders to make this possible. Then the operation of addition or subtraction is performed.

Example:

$0.423 + 1.562 + 0.0736 + 0.2 =$ ____

Solution:

Align decimal points

$$
\begin{array}{r}
0.4230 \\
+\ 1.5620 \\
+\ 0.0736 \\
0.2000 \\
\hline
2.2586
\end{array}
$$

Example:

$0.832 - 0.0357 =$ ____

Solution:

$$
\begin{array}{r}
0.8320 \\
-\ 0.0357 \\
\hline
0.7963
\end{array}
$$

Multiplying Decimals

When multiplying decimals, the decimal points do not have to be aligned. Rather, it is important to accurately position the decimal point in the product. To position the decimal in the product, the total number of digits to the right of the decimals in the numbers being multiplied must be equal to the number of digits to the right of the decimal in the product. This is best illustrated in the following examples:

Step 1: Multiply numbers without inserting decimal in the products.

Step 2: Sum the number of digits to the right of the decimal in all of the numbers being multiplied.

Step 3: Position the decimal in the product so the number of digits to the right of the decimal equals the total number of digits to the right of the decimal in the numbers multiplied (from Step 2).

Example:

0.056 x 0.032 = ____

Solution:

```
   0.056
   0.032
   -----
     112
    168
--------
0.001792
```

NOTE: Since 0.056 has three digits to the right of the decimal point, and 0.032 has three digits to the right of the decimal point, six digits must be to the right of the decimal point in the product. To have six digits in the product, zeros are inserted to the left of the computed digits.

To multiply a decimal by 10, move the decimal point one position to the right.

Example: 0.45 x 10 = 4.5. Similarly, when multiplying a decimal by 100, 1000, and 10,000, move the decimal point to the right the same number of zeros that are in the multiplier.

Example:

0.45 x 100 = 45
0.45 x 1000 = 450
0.45 x 10,000 = 4500

The reverse is true when multiplying by fractions of 10.

0.45 x 0.1 = 0.045
0.45 x 0.01 = 0.0045
0.45 x 0.001 = 0.00045
0.45 x 0.0001 = 0.000045

Dividing Decimals

When solving problems involving division of decimals, the following procedure should be applied.

Step 1: Write out the division problem.

Step 2: Move the decimal in the divisor to the right.

Step 3: Move the decimal in the dividend the same number of places to the right. Add zeros after the decimal in the dividend if necessary.

Step 4: Place the decimal point in the quotient directly above the decimal in the dividend.

Step 5: Divide the numbers.

Example:

$$3.00 \div 0.06 = \_\_\_\_$$

Solution:

Step 1: $.06\overline{)3.00}$

Step 2: $.06.\overline{)3.00}$

Step 3: $6.\overline{)300.}$

Step 4: $6.\overline{)300.}$

Step 5:

$$\begin{array}{r} 50. \\ 6\overline{)300.} \\ \underline{30} \\ 00 \end{array}$$

Rounding Off

When there is a remainder in division, the remainder may be written as a fraction or rounded off. When rounding off, the following rules should be applied:

Step 1: Observe the digit to the right of the digit being rounded off.

Step 2: If it is less than 5, drop the digit.
If the digit is 5 or higher, add 1 to the digit being rounded off.

Step 3: Write the new rounded number.

Example:

Round off the following number to two decimal places.

3.473

Solution:

Step 1: 3 is the number to the right of the 2nd decimal place.

Step 2: 3 is less than 5, so drop the digit.

Step 3: 3.47 is the number rounded to two decimal places.

Example:

Round off the following number to two decimal places.

6.238

Solution:

Step 1: 8 is the number to the right of the 2nd decimal place.

Step 2: 8 is greater than 5, so drop the 8 and add one to the number in the second decimal place (3 + 1 = 4).

Step 3: 6.24 is the number rounded to two decimal places.

Example:

Round off the following number to two decimal places.

6.2385

Solution:

Step 1: 8 is the number to the right of the 2nd decimal place.

Step 2: 8 is greater than 5, so drop the 8 and add one to number in the second decimal place (3 + 1 = 4).

Step 3: 6.24 is the number rounded to two decimal places.

Example:

Round off the following number to three decimal places.

6.2385

Solution:

Step 1: 5 is the number to the right of the 3rd decimal place.

Step 2: 5 is equal to 5, so drop the 5 and add one to the number in the third decimal place ($8 + 1 = 9$).

Step 3: 6.239 is the number rounded to three decimal places.

Example:

Divide 2.25 by 6 and round off the answer to 1 decimal place.

$$\frac{2.25}{6} = 0.375$$

Solution:

Step 1: 7 is the number to the right of the 1st decimal place.

Step 2: 7 is greater than 5, so drop the 7 and add one to the number in the first decimal place ($3 + 1 = 4$).

Step 3: 0.4 is .375 rounded to 1 decimal place.

Summary

The important information from this chapter is summarized below.

Decimals Summary

When using the decimal process:

Convert fractions to decimals by dividing the numerator by the denominator.

Convert decimals to fractions by writing the decimal in fraction format and reducing.

Align decimal points when adding or subtracting decimals.

Before dividing decimals, move the decimal in the divisor and dividend to the right by the same number of places.

When rounding, numbers less than 5 are dropped, and numbers 5 or greater increase the number immediately to the left by one.

SIGNED NUMBERS

This chapter covers the processes of addition, subtraction, division, and multiplication of signed numbers.

EO 1.7 APPLY one of the arithmetic operations of addition, subtraction, multiplication, and division using signed numbers.

Calculator Usage, Special Keys

Change Sign key

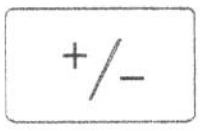

Pressing this key changes the sign of the number in the display. To enter a negative number, the number is entered as a positive number and then the change sign key is pressed to convert it to a negative. The display will show a "-" in front of the number.

Addition

Addition of signed numbers may be performed in any order. Begin with one number and count to the right if the other number is positive or count to the left if the other number is negative.

Example:

$$-2 + 3 = 0 - 2 + 3$$

Solution:

Begin with –2 and count 3 whole numbers to the right.

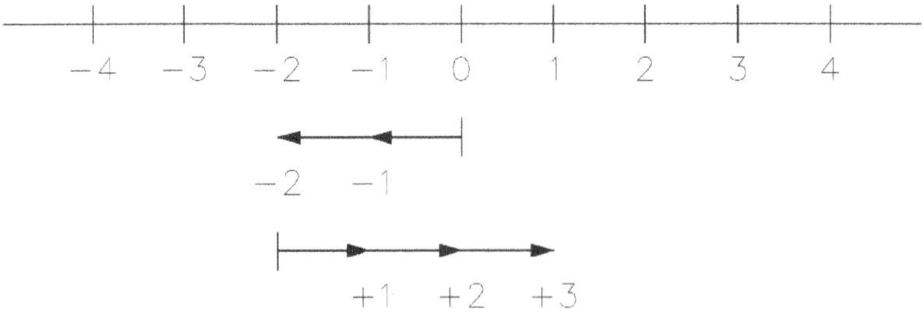

Therefore: -2 + 3 = 1

Example:

(-2) + 3 + 4 = 0 - 2 + 3 + 4

Solution:

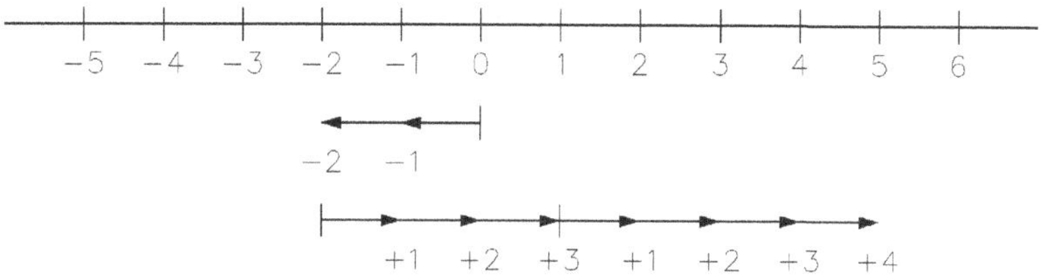

Therefore: (-2) + 3 + 4 = 5

Example:

(2) + (–4) = ____

Solution:

Begin with 2 and count 4 whole numbers to the left.

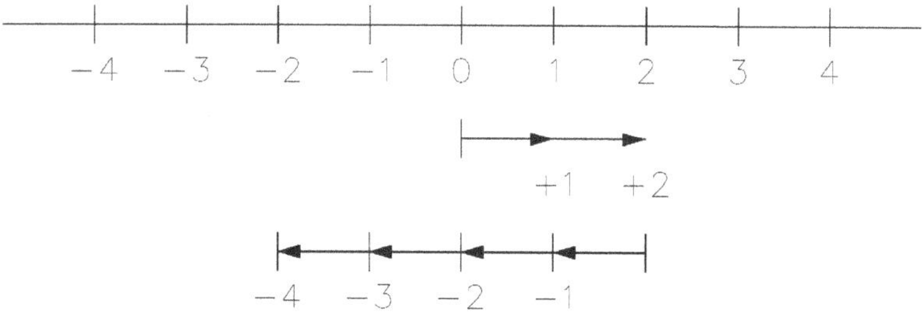

Therefore: (2) + (–4) = –2

Adding numbers with unlike signs may be accomplished by combining all positive numbers, then all negative numbers, and then subtracting.

Example:

$$10 + (-5) + 8 + (-7) + 5 + (-18) = \_\_\_\_$$

Solution:

$$+10 - 5 + 8 - 7 + 5 - 18 =$$

$$+10 + 8 + 5 - 18 - 7 - 5 =$$

$$+23 - 30 = -7$$

Subtraction

Subtraction of signed numbers may be regarded as the addition of numbers of the opposite signs. To subtract signed numbers, reverse the sign of the subtrahend (the second number) and add.

For example, one could treat his incomes for a given month as positive numbers and his bills as negative numbers. The difference of the two is his increase in cash balance. Suppose he buys a window for $40. This gives a bill of $40 and adds as negative $40 to his cash balance. Now suppose he returns this window to the store and the manager tears up his bill, subtracting the -$40. This is equivalent of adding +$40 to his cash balance.

Example:

$$a - b \quad = \quad a + (-b)$$

Solution:

$$(+3) - (+5) \quad = \quad (+3) + (-5) \quad = -2$$

$$(-4) - (-1) \quad = \quad (-4) + (+1) \quad = -3$$

$$(-5) - (+8) \quad = \quad (-5) + (-8) \quad = -13$$

$$(+7) - (-2) \quad = \quad (+7) + (+2) \quad = +9$$

Multiplication

Multiplication of signed numbers may be performed by using the following rules:

The product of any two numbers with like signs is positive:
$(+)(+) = (+)$ or $(-)(-) = (+)$.

The product of any two numbers with unlike signs is negative:
$(+)(-) = (-)$ or $(-)(+) = (-)$.

The product is negative if there is an odd number of negatives.
The product is positive if there is an even number of negatives.

Example:

$(+3)(+3) = +9$
$(-2)\ (+4) = -8$
$(-1)\ (-2)\ (+1)\ (-2) = -4$
$(-2)\ (+2)\ (+2)\ (-2) = +16$

Zero times any number equals zero.
Multiplying by -1 is the equivalent of changing the sign.

Division

Division of signed numbers may be performed using the following rules:

Rule 1: The quotient of any two numbers with like signs is positive:
$(+)/(+) = (+)$ or $(-)/(-) = (+)$

Rule 2: The quotient of any two numbers with unlike signs is negative:
$(+)/(-) = (-)$ or $(-)/(+) = (-)$

Rule 3: Zero divided by any number not equal to zero is zero.

Examples:

a) $\frac{0}{-5} = 0$ Apply rule 3.

b) $\frac{-3}{-1} = +3$ Apply rule 1.

c) $\frac{-4}{+2} = -2$ Apply rule 2.

Summary

The important information from this chapter is summarized below.

Signed Numbers Summary

When using signed numbers:

Adding a negative number is the same as subtracting a positive number.

Subtracting a negative number is the same as adding a positive number.

A product is negative if there is an odd number of negatives.

A product is positive if there is an even number of negatives.

Division of two numbers with like signs results in a positive answer.

Division of two numbers with unlike signs results in a negative answer.

SIGNIFICANT DIGITS

This chapter presents the concept of significant digits and the application of significant digits in a calculation.

EO 1.8 **DETERMINE the number of significant digits in a given number.**

EO 1.9 **Given a formula, CALCULATE the answer with the appropriate number of significant digits.**

Calculator Usage, Special Keys

Most calculators can be set up to display a fixed number of decimal places. In doing so, the calculator continues to perform all of its internal calculations using its maximum number of places, but rounds the displayed number to the specified number of places.

INV key

To fix the decimal place press the INV key and the number of the decimal places desired. For example, to display 2 decimal places, enter INV 2.

Significant Digits

When numbers are used to represent a measured physical quantity, there is uncertainty associated with them. In performing arithmetic operations with these numbers, this uncertainty must be taken into account. For example, an automobile odometer measures distance to the nearest 1/10 of a mile. How can a distance measured on an odometer be added to a distance measured by a survey which is known to be exact to the nearest 1/1000 of a mile? In order to take this uncertainty into account, we have to realize that we can be only as precise as the least precise number. Therefore, the number of significant digits must be determined.

Suppose the example above is used, and one adds 3.872 miles determined by survey to 2.2 miles obtained from an automobile odometer. This would sum to 3.872 + 2.2 = 6.072 miles, but the last two digits are not reliable. Thus the answer is rounded to 6.1 miles. Since all we know about the 2.2 miles is that it is more than 2.1 and less than 2.3, we certainly don't know the sum to any better accuracy. A single digit to the right is written to denote this accuracy.

Both the precision of numbers and the number of significant digits they contain must be considered in performing arithmetic operations using numbers which represent measurement. To determine the number of significant digits, the following rules must be applied:

Rule 1: The left-most non-zero digit is called the most significant digit.

Rule 2: The right-most non-zero digit is called the least significant digit except when there is a decimal point in the number, in which case the right-most digit, even if it is zero, is called the least significant digit.

Rule 3: The number of significant digits is then determined by counting the digits from the least significant to the most significant.

Example:

In the number 3270, 3 is the most significant digit, and 7 is the least significant digit.

Example:

In the number 27.620, 2 is the most significant digit, and 0 is the least significant digit.

When adding or subtracting numbers which represent measurements, the right-most significant digit in the sum is in the same position as the left-most least significant digit in the numbers added or subtracted.

Example:

15.62 psig + 12.3 psig = 27.9 psig

Example:

401.1 + 50 = 450

Example:

401.1 + 50.0 = 451.1

When multiplying or dividing numbers that represent measurements, the product or quotient has the same number of significant digits as the multiplied or divided number with the least number of significant digits.

Example:

3.25 inches x 2.5 inches = 8.1 inches squared

Summary

The important information from this chapter is summarized below.

Significant Digits Summary

Significant digits are determined by counting the number of digits from the most significant digit to the least significant digit.

When adding or subtracting numbers which represent measurements, the right-most significant digit in the sum is in the same position as the left-most significant digit in the numbers added or subtracted.

When multiplying or dividing numbers that represent measurements, the product or quotient has the same number of significant digits as the multiplied or divided number with the least number of significant digits.

PERCENTAGES

This chapter covers the conversion between percents, decimals, and fractions.

EO 1.10 CONVERT between percents, decimals, and fractions.

EO 1.11 CALCULATE the percent differential.

A special application of proper fractions is the use of percentage. When speaking of a 30% raise in pay, one is actually indicating a fractional part of a whole, 30/100. The word percent means "hundredth;" thus, 30% is based on the whole value being 100%. However, to perform arithmetic operations, the 30% expression is represented as a decimal equivalent (0.30) rather than using the % form.

Calculator Usage, Special Keys

Percent Key

%

When pressed, the percent key divides the displayed number by 100.

Changing Decimals to Percent

Any number written as a decimal may be written as a percent. To write a decimal as a percent, multiply the decimal by 100, and add the percent symbol.

Example:

Change 0.35 to percent.
0.35 x 100 = 35%

Example:

Change 0.0125 to percent.
0.0125 x 100 = 1.25%

Example:

Change 2.7 to percent.
$2.7 \times 100 = 270\%$

Changing Common Fractions and Whole Numbers to Percent

When changing common fractions to percent, convert the fraction to a decimal, then multiply by 100 and add the percent symbol.

Example:

Change $\frac{3}{5}$ to a percent
$0.6 \times 100 = 60\%$

When changing a whole number to a percent, multiply by 100 and add the percent symbol.

Example:

Change 10 to percent
$10 \times 100 = 1000\%$

Percents are usually 100% or less. Percents are most often used to describe a fraction, but can be used to show values greater than 1(100%). Examples are 110%, 200%, etc.

Changing a Percent to a Decimal

Any number written as a percent may be written as a decimal. To change a percent to a decimal, drop the percent symbol and divide by 100.

Example:

Express 33.5% in decimal form.
$$\frac{33.5}{100} = 0.335$$

Express 3.35% in decimal form.
$$\frac{3.35}{100} = 0.0335$$

Express 1200% in decimal form.

$$\frac{1200}{100} = 12$$

Percent Differential

Percent differentials are used to provide a means of comparing changes in quantities or amounts. Percent differentials express the relationship between some initial condition and another specified condition.

The method of calculating percent differential involves the following:

Step 1: Subtract the original value from the present value.

Step 2: Divide by the original value.

Step 3: Multiply by 100.

Step 4: Add the percent symbol (%).

Example:

A tank initially contains 50 gallons of water. Five gallons are drained out. By what percent is the amount of water in the tank reduced?

Solution:

Step 1: The difference between initial and final is given in the problem: 5 gallons.

Step 2: $\frac{5}{50} = 0.1$

Step 3: 0.1 x 100 = 10% Five gallons represents 10% of the original 50 gals that were in the tank.

Ratio

Two numbers may be compared by expressing the relative size as the quotient of one number divided by the other and is called a **ratio**. Ratios are simplified fractions written with a colon (:) instead of a division bar or slash.

Example:

One day Eric paid $700 for a stereo and Scott paid $600 for the same stereo. Compare the amount that Eric paid to the amount that Scott paid, using ratios.

Solution:

Step 1: Divide the numbers to be compared. In this example the amount paid by Scott is being compared to the amount paid by Eric. The amount paid by Eric is divided by the amount paid by Scott = $\frac{700}{600}$.

Step 2: Simplifying this expression, both 700 and 600 can be divided by 100.

Step 3: Expressing this fraction as a ratio:

$$\frac{\text{Eric's price}}{\text{Scott's price}} = \frac{7}{6} \text{ or Eric's price : Scott's price} = 7:6$$

Example:

If one yard equals three feet, what is the ratio of yards to feet?

Solution:

Step 1: 1 yd./ 3 ft.

Step 2: $\frac{1}{3}$ is already in simplest terms

Step 3: $\frac{\text{yards}}{\text{feet}} = \frac{1}{3}$ or yards : feet = 1:3

Summary

Pertinent information concerning percentages and ratios is summarized below.

Percentages and Ratios Summary

Change decimals to percents by multiplying by 100 and adding the percent symbol.

Change fractions to percents by first changing the fraction into a decimal. Then change the decimal to a percent.

Compute percent differential by dividing the difference by the original value, multiplying by 100, and adding the percent symbol.

Ratios are fractions written with a colon instead of a division bar or slash.

EXPONENTS

This chapter covers the addition, subtraction, multiplication, and division of numbers with exponents.

EO 1.12 APPLY one of the arithmetic operations of addition, subtraction, multiplication, and division using exponential numbers.

Calculator Usage, Special Keys

Exponent key

y^x

Raising a number to an exponent requires the y^x key to be pressed twice. First, the base number is entered and the y^x key is pressed; this enters the base number (y). Next, the exponent number is pressed and the y^x key is pressed; this enters the exponent and tells the calculator to complete the calculation. The calculator will display the value.

x squared key

x^2

Pressing this key squares the displayed number. This key will save time over using the y^x key.

Exponents

The product $a \times a \times a \times a$ can be written as a^4, where 4 is called the exponent of a or power to which a is raised. In this notation, a is often called the base.

Examples:

$$a^4 = a \cdot a \cdot a \cdot a$$
$$5^3 = 5 \cdot 5 \cdot 5$$
$$(a + b)^5 = (a+b) \cdot (a+b) \cdot (a+b) \cdot (a+b) \cdot (a+b)$$

When an exponent is not written, it is assumed to be 1. For example, $a^1 = a$. An exponent applies only to the quantity immediately to the left and below it. For example, in $3 + (-2)^3$ the base is -2, but in $3 - 2^3$ the base is 2.

Basic Rules for Exponents

The following rules are applied to exponents.

Rule 1: To multiply numbers with the same base, add the exponents and keep the base the same.

$$a^m a^n = a^{m+n}$$

Example:

$2^2 \times 2^3 = (2 \times 2) \times (2 \times 2 \times 2) = 2 \times 2 \times 2 \times 2 \times 2 = 2^5$

Rule 2: When raising a power of a number to a power, multiply the exponents and keep the base the same.

$$(a^m)^n = a^{mn}$$

Example:

$(a^2)^3 = (a \times a) \times (a \times a) \times (a \times a) = a^6$

that is, you multiply $(a \times a)$ three times. Similarly, for $(a^m)^n$, one multiplies (a^m) n times. There are m values of a in each parenthesis multiplied by n parenthesis or m x n values of a to multiply.

Thus, $(a^m)^n = a^{mn}$

Rule 3: When dividing two exponential numbers, subtract the powers.

$$\frac{a^m}{a^n} = a^{m-n}$$

Example:

$$\frac{a^5}{a^2} = \frac{a \times a \times a \times a \times a}{a \times a} = \frac{\cancel{a}}{\cancel{a}} \times \frac{\cancel{a}}{\cancel{a}} \times a \times a \times a = a^3$$

Rule 4: Any exponential number divided by itself is equal to one.

$$\frac{a^n}{a^n} = 1$$

Rule 5: To raise a product to a power, raise each factor to that power.

$$(ab)^n = a^n b^n$$

This arises from the associative law for multiplication, that is, order of multiplication does not alter the product.

Example:

$$(ab)^2 = (a \times b) \times (a \times b) = (a \times a)\ (b \times b) = a^2 \times b^2$$

If doubt exists in the student's mind, try multiplying $(2 \times 3)^2$ out in different orders. All orders will yield 36.

Rule 6: To raise a quotient to a power, raise both the numerator and denominator to that power.

$$\left(\frac{a}{b}\right)^n = \frac{a^n}{b^n}$$

Example:

To demonstrate this, consider $\left(\frac{3}{2}\right)^2 = 1.5^2 = 2.25 = 2\frac{1}{4} = \frac{9}{4}$

But $\frac{3^2}{2^2} = \frac{9}{4}$, the same value.

Zero Exponents

Using the rule for exponents (Rule 4) to evaluate a^n/a^n, then

$$\frac{a^n}{a^n} = 1$$

This interpretation is consistent with the rule $a^n/a^n = a^{n-n} = a^0$. Therefore, $a^0 = 1$ when a is not equal to 0. Any number to the zero power equals one.

Example:

$3^0 = 1$
$(b^2+2)^0 = 1$

Negative Exponents

The rules for positive exponents apply to negative exponents. a^{-n} is defined as follows:

$$a^{-n} = a^{n} = \frac{1}{a^n}$$

$$a^n = \frac{1}{a^{-n}}$$

For example, $a^5/a^2 = a^{5-2}$ as shown earlier. If $\frac{1}{a^2}$ is written as a^{-2}, and the rules for multiplication are applied to this, $a^5 \times a^{-2} = a^{5-2} = a^3$. Thus, writing $\frac{1}{a^n}$ as a^{-n} and applying the rules for multiplication yields the same results as $\frac{1}{a^n}$ and applying the rules of division.

Examples:

$$c^{-2} = \frac{1}{c^2}$$

$$x^3 = \frac{1}{x^{-3}}$$

Fractional Exponents

Fractional exponents are defined as follows, $a^{\frac{1}{m}} \equiv \sqrt[n]{a}$. This permits manipulations with numbers with fractional exponents to be treated using the laws expressed earlier for integers. For example,

$$8^{\frac{1}{3}} \equiv \sqrt[3]{8} = 2 \qquad \text{since } 2 \times 2 \times 2 = 8$$

Taking the statement $8^{\frac{1}{3}} = 2$ and cubing both sides, $\left(8^{\frac{1}{3}}\right)^3 = 2^3$. But $(a^m)^n = a^{m \times n}$ so $\left(8^{\frac{1}{3}}\right)^3 = 8^1 = 8$ which agrees with $2^3 = 8$ for the right-hand side of the equality.

A number such as $8^{\frac{2}{3}}$ can be written $\left(8^{\frac{1}{3}}\right)^2 = 2^2 = 4$ or alternately as $(8^2)^{\frac{1}{3}} = (64)^{\frac{1}{3}} = 4$ since 4 x 4 x 4 = 64; that is, 4 is the cube root of 64.

Examples:

$$\left(a^{\frac{1}{3}}\right) \cdot \left(a^{\frac{2}{3}}\right) = a^{\left(\frac{1}{3} + \frac{2}{3}\right)} = a^1 = a$$

$$\frac{b^{\frac{1}{4}}}{b^{\frac{1}{2}}} = b^{\left(\frac{1}{4} - \frac{1}{2}\right)} = b^{-\frac{1}{2}} = \frac{1}{b^2}$$

$$\left(d^{\frac{1}{3}}\right)^9 = d^{\left(\frac{1}{3} \times 9\right)} = d^3$$

Summary

Pertinent information concerning exponents is summarized below.

Exponents Summary

$\text{Base}^{\text{Exponent}} = \text{Product}$

Rule 1: To multiply numbers with the same base, add the exponents and keep the base the same.

$$a^m a^n = a^{m+n}$$

Rule 2: When raising a power of a number to a power, multiply the exponents and keep the base the same.

$$(a^m)^n = a^{mn}$$

Rule 3: When dividing two exponential numbers, subtract the powers.

$$a^m/a^n = a^{m-n}$$

Rule 4: Any exponential number divided by itself is equal to one.

$$a^n/a^n = 1$$

Rule 5: To raise a product to a power, raise each factor to that power.

$$(ab)^n = a^n b^n$$

Rule 6: To raise a quotient to a power, raise both the numerator and denominator to that power.

$$(a/b)^n = a^n/b^n$$

Any number to the zero power equals one.

The rules for positive exponents apply to negative exponents.

The rules for integer exponents apply to fractional exponents.

SCIENTIFIC NOTATION

This chapter covers the addition, subtraction, multiplication, and division of numbers in scientific notation.

EO 1.13 **Given the data, CONVERT integers into scientific notation and scientific notation into integers.**

EO 1.14 **APPLY one of the arithmetic operations of addition, subtraction, multiplication, and division to numbers using scientific notation.**

Calculator Usage

Scientific Notation key

EE

If pressed after a number is entered on the display, the EE key will convert the number into scientific notation. If a number is to be entered in scientific notation into the calculator, pressing the EE key tells the calculator the next entered numbers are the exponential values.

Scientists, engineers, operators, and technicians use scientific notation when working with very large and very small numbers. The speed of light is 29,900,000,000 centimeters per second; the mass of an electron is 0.000549 atomic mass units. It is easier to express these numbers in a shorter way called scientific notation, thus avoiding the writing of many zeros and transposition errors.

$$29{,}900{,}000{,}000 = 2.99 \times 10^{10}$$
$$0.000549 = 5.49 \times 10^{-4}$$

Writing Numbers in Scientific Notation

To transform numbers from decimal form to scientific notation, it must be remembered that the laws of exponents form the basis for calculations using powers.

Using the results of the previous chapter, the following whole numbers and decimals can be expressed as powers of 10:

$1 = 10^0$	$0.1 = 1/10 = 10^{-1}$
$10 = 10^1$	$0.01 = 1/100 = 10^{-2}$
$100 = 10^2$	$0.001 = 1/1000 = 10^{-3}$
$1000 = 10^3$	
$10{,}000 = 10^4$	

A number N is in scientific notation when it is expressed as the product of a decimal number between 1 and 10 and some integer power of 10.

$$N = a \times 10^n \text{ where } 1 \leq a < 10 \text{ and } n \text{ is an integer.}$$

The steps for converting to scientific notation are as follows:

Step 1: Place the decimal immediately to the right of the left-most non-zero number.

Step 2: Count the number of digits between the old and new decimal point.

Step 3: If the decimal is shifted to the left, the exponent is positive. If the decimal is shifted to the right, the exponent is negative.

Let us examine the logic of this. Consider as an example the number 3750. The number will not be changed if it is multiplied by 1000 and divided by 1000 (the net effect is to multiply it by one). Then,

$$\frac{3750}{1000} \times 1000 = 3.750 \times 1000 = 3.750 \times 10^3$$

There is a division by 10 for each space the decimal point is moved to the left, which is compensated for by multiplying by 10. Similarly, for a number such as .0037, we multiply the number by 10 for each space the decimal point is moved to the right. Thus, the number must be divided by 10 for each space.

$$.0037 = .0\,0\,3.7 \times \frac{1}{10} \times \frac{1}{10} \times \frac{1}{10} = \frac{3.7}{10^3} = 3.7 \times 10^{-3}$$

Example 1:

Circulating water flows at 440,000 gallons per minute. Express this number in scientific notation.

Solution:

440,000 becomes 4.4×10^n

4 4 0 0 0 0.
5 4 3 2 1

$n = +5$ because the decimal is shifted five places to the left.

$440{,}000 = 4.4 \times 10^5$

Example 2:

Express 0.0000247 in scientific notation.

Solution:

0. 0 0 0 0 2 4 7
1 2 3 4 5

$n = -5$ because the decimal is shifted five places to the right.

$0.0000247 = 2.47 \times 10^{-5}$

Example 3:

Express 34.2 in scientific notation.

Solution:

3 4. 2
1

$n = 1$ because the decimal is shifted one place to the left.

$34.2 = 3.42 \times 10^1$

Converting Scientific Notation to Integers

Often, numbers in scientific notation need to be put in integer form.

To convert scientific notation to integers:

Step 1: Write decimal number.

Step 2: Move the decimal the number of places specified by the power of ten: to the right if positive, to the left if negative. Add zeros if necessary.

Step 3: Rewrite the number in integer form.

Example:

Convert 4.4×10^3 to integer form.

Solution:

Step 1: 4.4

Step 2: $4.\underbrace{4}_{1}\underbrace{0}_{2}\underbrace{0}_{3}$

Step 3: $4.4 \times 10^3 = 4400$

Addition

In order to add two or more numbers using scientific notation, the following three steps must be used.

Step 1: Change all addends to have the same power of ten by moving the decimal point (that is, change all lower powers of ten to the highest power).

Step 2: Add the decimal numbers of the addends and keep the common power of ten.

Step 3: If necessary, rewrite the decimal with a single number to the left of the decimal point.

For example, for $3.5 \times 10^3 + 5 \times 10^2$ you are asked to add 3.5 thousands to 5 hundreds. Converting 3.5 thousands to 35 hundreds ($3.5 \times 10^3 = 35 \times 10^2$) we obtain 35 hundreds + 5 hundreds = 40 hundreds or $3.5 \times 10^3 = 35 \times 10^2 + 5 \times 10^2 = 4 \times 10^3$. The student should do the same problem by converting the 5×10^2 to thousands and then adding.

Example:

Add $(9.24 \times 10^4) + (8.3 \times 10^3)$

Solution:

Step 1: $9.24 \times 10^4 = 9.24 \times 10^4$
$8.3 \times 10^3 = 0.83 \times 10^4$

Step 2: 9.24×10^4
$\underline{+0.83 \times 10^4}$

Step 3: $10.07 \times 10^4 = 1.007 \times 10^5$

Subtraction

In order to subtract two numbers in scientific notation, the steps listed below must be followed.

Step 1: As in addition, change all addends to have the same power of ten.

Step 2: Subtract one digit from the other and keep the power of ten.

Step 3: If necessary, rewrite the decimal with a single number to the left of the decimal point.

Example:

Subtract $(3.27 \times 10^4) - (2 \times 10^3)$

Solution:

Step 1: $3.27 \times 10^4 = 3.27 \times 10^4$
$2.00 \times 10^3 = 0.20 \times 10^4$

Step 2: 3.27×10^4
$\underline{-0.20 \times 10^4}$

Step 3: 3.07×10^4

Multiplication

When multiplying two or more numbers in scientific notation, the following steps must be used.

Step 1: Multiply the decimal numbers and obtain the product.

Step 2: Multiply the powers of ten together by adding the exponents.

Step 3: Put the product in single-digit scientific notation.

Step 4: If necessary, rewrite decimal with a single number to the left of the decimal point.

Example:

Multiply $(3 \times 10^3)(5 \times 10^{-2})$

Solution:

Step 1: $3 \times 5 = 15$

Step 2: $10^3 \times 10^{-2} = 10^{3+-2} = 10^1$

Step 3: The product is: 15×10^1

Step 4: $= 1.5 \times 10^2$

Division

Follow the steps listed below when dividing numbers in scientific notation.

Step 1: Divide one decimal into the other.

Step 2: Divide one power of ten into the other by subtracting the exponents.

Step 3: Put product in single-digit scientific notation.

Step 4: If necessary, rewrite decimal with a single number to the left of the decimal point.

Example:

$(1 \times 10^6) \div 5 \times 10^4 =$

Solution:

Step 1: $\frac{1}{5} - 0.2$

Step 2: $\frac{10^6}{10^4} = 10^{(6-4)} = 10^2$

Step 3: 0.2×10^2

Step 4: 2.0×10^1

Summary

Pertinent information concerning scientific notation is summarized below.

Scientific Notation Summary

When changing from integer form to scientific notation:

- If the decimal is shifted left, the exponent is positive.
- If the decimal is shifted right, the exponent is negative.

When adding or subtracting numbers in scientific notation, change both numbers to the same power of ten by moving the decimal point. Add or subtract the decimal numbers, and keep the power of ten. Rewrite if necessary.

To multiply two numbers in scientific notation, multiply decimal numbers and add exponents. Rewrite if necessary.

To divide two numbers in scientific notation, divide decimal numbers and subtract exponents. Rewrite if necessary.

RADICALS

This chapter covers the addition, subtraction, multiplication, and division of radicals.

EO 1.15 CALCULATE the numerical value of numbers in radical form.

Calculator Usage, Special Keys

The exponent key can be used for radicals if the exponent is entered in decimal form.

Exponent key

y^x

Raising a number to an exponent requires the y^x key to be pressed twice. First, the base number is entered and the y^x key is pressed. This enters the base number (y). Next, the exponent number is entered and the y^x key is pressed. This enters the exponent and tells the calculator to complete the calculation. The calculator will display the value.

Square-root key

$\sqrt{x}$

Pressing this key takes the square root of the displayed number.

The Radical

A previous chapter explained how to raise a number to a power. The inverse of this operation is called extracting a root. For any positive integer n, a number x is the nth root of the number a if it satisfies $x^n = a$. For example, since $2^5 = 32$, 2 is the fifth root of 32.

To indicate the nth root of a, the expression $a^{1/n}$ is often used. The symbol $\sqrt{\ }$ is called the radical sign, and the nth root of a can also be shown as $\sqrt[n]{a}$. The letter a is the radicand, and n is the index. The index 2 is generally omitted for square roots.

Example:

$$\sqrt{4} = 2$$

$$\sqrt[3]{27} = 3$$

$$\sqrt[4]{x^4} = x$$

Simplifying Radicals

An expression having radicals is in simplest form when:

- The index cannot be reduced.
- The radicand is simplified.
- No radicals are in the denominator.

There are four rules of radicals that will be useful in simplifying them.

Rule 1: $\left(\sqrt[n]{a}\right)^n = \sqrt[n]{a^n} = a$

Rule 2: $\sqrt[n]{ab} = \sqrt[n]{a}\,\sqrt[n]{b}$

Rule 3: $\sqrt[n]{\dfrac{a}{b}} = \dfrac{\left(\sqrt[n]{a}\right)}{\left(\sqrt[n]{b}\right)}$

Rule 4: $\sqrt[n]{-a} = -\sqrt[n]{a}$, when n is odd.

Examples: $\sqrt{10^2} = 10$

$\left(\sqrt[3]{26}\right)^3 = 26$

$\sqrt{27} = \sqrt{9 \cdot 3} = \sqrt{9}\,\sqrt{3} = 3\sqrt{3}$

$\sqrt[3]{-54} = \sqrt[3]{(-27)(2)} = \left(\sqrt[3]{-27}\right)\left(\sqrt[3]{2}\right) = -3\,\sqrt[3]{2}$

When a radical sign exists in the denominator, it is desirable to remove the radical. This is done by multiplying both the numerator and denominator by the radical and simplifying.

Example: $\dfrac{3}{\sqrt{5}} = \dfrac{3}{\sqrt{5}} \cdot \dfrac{\sqrt{5}}{\sqrt{5}} = \dfrac{3\sqrt{5}}{5}$

Addition and Subtraction

Addition and subtraction of radicals may be accomplished with radicals showing the same radicand and the same index. Add or subtract similar radicals using the distributive law.

Examples: $3\sqrt{ab} + 2\sqrt{ab} = (3 + 2)\sqrt{ab} = 5\sqrt{ab}$

$$7\sqrt{5} - 3\sqrt{5} = (7 - 3)\sqrt{5} = 4\sqrt{5}$$

Multiplication

Multiplication of radicals having the same index may be accomplished by applying the rule used in simplification: $\sqrt[n]{ab} = \sqrt[n]{a} \cdot \sqrt[n]{b}$

Examples: $\sqrt[3]{3x^4} \cdot \sqrt[3]{9x^2} = \sqrt[3]{27x^6} = 3x^2$

$$\sqrt{xy} \cdot \sqrt{3x} = \sqrt{3x^2y} = x\sqrt{3y}$$

$$\sqrt[4]{\frac{16}{3}} \cdot \sqrt[4]{\frac{2}{27}} = \sqrt[4]{\frac{32}{81}} = \frac{\sqrt[4]{32}}{\sqrt[4]{81}} = \frac{\sqrt[4]{2 \cdot 16}}{\sqrt[4]{81}} = \frac{2\sqrt[4]{2}}{3}$$

Division

Division of radicals having the same index, but not necessarily the same radicand, may be performed by using the following rule and simplifying.

Examples: $\dfrac{\sqrt[n]{a}}{\sqrt[n]{b}} = \sqrt[n]{\dfrac{a}{b}}$

$$\frac{\sqrt{18}}{\sqrt{2}} = \sqrt{\frac{18}{2}} = \sqrt{9} = 3$$

Dissimilar Radicals

Often, dissimilar radicals may be combined after they are simplified.

Example: $\sqrt[4]{81x^2} + \sqrt{x} - \sqrt[6]{64x^3}$

$= 3\sqrt{x} + \sqrt{x} - 2\sqrt{x}$

$= (3+1-2)\ \sqrt{x} = 2\sqrt{x}$

Changing Radicals to Exponents

This chapter has covered solving radicals and then converting them into exponential form. It is much easier to convert radicals to exponential form and then perform the indicated operation.

The expression $\sqrt[3]{4}$ can be written with a fractional exponent as $4^{1/3}$. Note that this meets the condition $\left(4^{\frac{1}{3}}\right)^3 = 4$, that is, the cube root of 4 cubed equals 4. This can be expressed in the following algebraic form:

$$a^{1/n} = \sqrt[n]{a}$$

The above definition is expressed in more general terms as follows:

$$a^{m/n} = \left(\sqrt[n]{a}\right)^m = \sqrt[n]{a^m}$$

Example 1:

Express the following in exponential form.

$\sqrt[3]{27^2} = 27^{2/3}$

$\sqrt{2} = 2^{1/2}$

Example 2:

Solve the following by first converting to exponential form.

$\sqrt{27} \cdot \sqrt[3]{27} = 27^{1/2} \cdot 27^{1/3} = 27^{5/6}$

but $27 = 3^3$

substituting: $27^{5/6} = (3^3)^{5/6} = 3^{5/2}$

Changing Exponents to Radicals

How to convert radicals into exponential form has been explained. Sometimes however, it is necessary or convenient to convert exponents to radicals. Recognizing that an exponent is the equivalent of the nth root is useful to help comprehend an expression.

The expression $5^{1/3}$ can be written as $\sqrt[3]{5}$. It is algebraically expressed as:

$$a^{\frac{1}{m}} = \sqrt[m]{a}$$

The above definition can be more generally described as:

$$\sqrt[n]{a^m} = (a^m)^{\frac{1}{n}} = a^{\frac{m}{n}}$$

and

$$\left(\sqrt[n]{b}\right)^m = \left(b^{\frac{1}{n}}\right)^m = b^{\frac{m}{n}}$$

Examples: $15^{2/3} = \sqrt[3]{15^2}$

$16^{1/2} = \sqrt[2]{16} = 4$

Summary

Pertinent information concerning radicals is summarized below.

Radicals Summary

- $\left(\sqrt[n]{a}\right)^n = \sqrt[n]{a^n} = a$ Used in simplification
- $\sqrt[n]{ab} = \sqrt[n]{a}\,\sqrt[n]{b}$ Used in simplification and multiplication
- $\sqrt[n]{\frac{a}{b}} = \frac{\left(\sqrt[n]{a}\right)}{\left(\sqrt[n]{b}\right)}$ Used in simplification and division
- $\sqrt[n]{a} = a^{1/n}$ Used to change radicals to exponents and exponents to radicals

Intentionally Left Blank

Appendix A
TI-30 Keyboard

Review of Introductory Mathematics

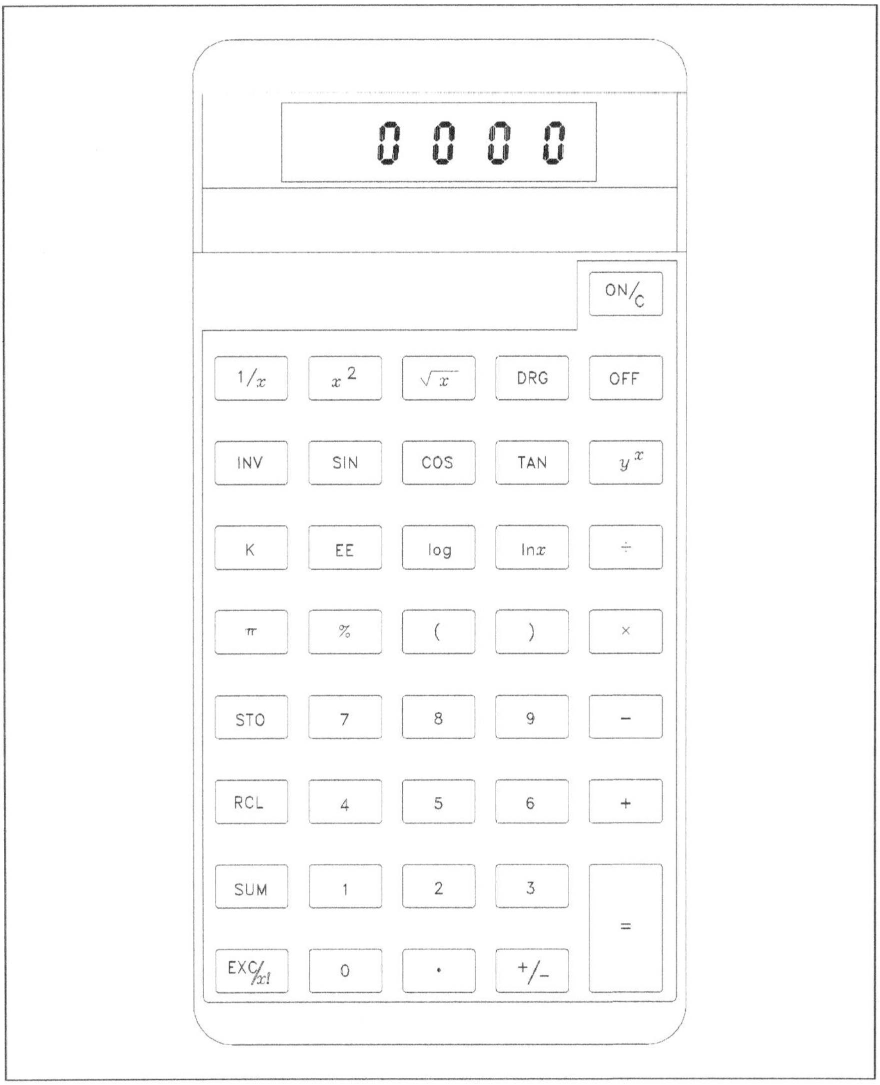

Figure A-1 TI-30 Keyboard Layout

Intentionally Left Blank

Department of Energy
Fundamentals Handbook

MATHEMATICS
Module 2
Algebra

TABLE OF CONTENTS

TABLE OF CONTENTS (Cont)

LIST OF FIGURES

LIST OF TABLES

REFERENCES

- Dolciani, Mary P., et al., Algebra Structure and Method Book 1, Atlanta: Houghton-Mifflin, 1979.
- Naval Education and Training Command, Mathematics, Volume 1, NAVEDTRA 10069-D1, Washington, D.C.: Naval Education and Training Program Development Center, 1985.
- Science and Fundamental Engineering, Windsor, CT: Combustion Engineering, Inc., 1985.
- Academic Program For Nuclear Power Plant Personnel, Volume 1, Columbia, MD: General Physics Corporation, Library of Congress Card #A 326517, 1982.

TERMINAL OBJECTIVE

1.0 Given a calculator and a list of formulas, **APPLY** the laws of algebra to solve for unknown values.

ENABLING OBJECTIVES

1.1 Given an equation, **DETERMINE** the governing algebraic law from the following:
- a. Commutative law
- b. Associative law
- c. Distributive law

1.2 **SOLVE** for the unknown given a linear equation.

1.3 **APPLY** the quadratic formula to solve for an unknown.

1.4 Given simultaneous equations, **SOLVE** for the unknowns.

1.5 Given a word problem, **WRITE** equations and **SOLVE** for the unknown.

1.6 **STATE** the definition of a logarithm.

1.7 **CALCULATE** the logarithm of a number.

1.8 **STATE** the definition of the following terms:
- a. Ordinate
- b. Abscissa

1.9 Given a table of data, **PLOT** the data points on a cartesian coordinate graph.

1.10 Given a table of data, **PLOT** the data points on a logarithmic coordinate graph.

1.11 Given a table of data, **PLOT** the data points on the appropriate graphing system to obtain the specified curve.

ENABLING OBJECTIVES (Cont)

1.12 **OBTAIN** data from a given graph.

1.13 Given the data, **SOLVE** for the unknown using a nomograph.

1.14 **STATE** the definition of the following terms:
   a. Slope
   b. Intercept

1.15 Given the equation, **CALCULATE** the slope of a line.

1.16 Given the graph, **DETERMINE** the slope of a line.

1.17 Given a graph, **SOLVE** for the unknown using extrapolation.

1.18 Given a graph, **SOLVE** for the unknown using interpolation.

Intentionally Left Blank

ALGEBRAIC LAWS

This chapter covers the laws used for solving algebraic equations.

EO 1.1 Given an equation, DETERMINE the governing algebraic law from the following:

a. Commutative law
b. Associative law
c. Distributive law

Most of the work in basic mathematics completed by DOE facility personnel involves real numbers, as mentioned in the last section. As a result, one should be very familiar with the basic laws that govern the use of real numbers. Most of these laws are covered under the general area called Algebra.

Algebraic Laws

Many operations on real numbers are based on the commutative, associative, and distributive laws. The effective use of these laws is important. These laws will be stated in written form as well as algebraic form, where letters or symbols are used to represent an unknown number.

The **commutative laws** indicate that numbers can be added or multiplied in any order.

Commutative Law of Addition: $a + b = b + a$

Commutative Law of Multiplication: $a(b) = b(a)$

The **associative laws** state that in addition or multiplication, numbers can be grouped in any order.

Associative Law of Addition: $a+(b+c) = (a+b)+c$

Associative Law of Multiplication: $a(bc) = (ab)c$

The **distributive laws** involve both addition and multiplication and state the following.

Distributive law: $a(b + c) = ab + ac$

Distributive law: $(a + b)c = ac + bc$

The following list of axioms pertains to the real number system where a, b, and c represent any real numbers. These properties must be true for the algebraic laws to apply.

Closure Properties	1.	$a + b$ is a real number
	2.	ab is a real number
Identity Properties	3.	$a + 0 = a$
	4.	$a(1) = a$
Inverse Properties	5.	For every real number, a, there exists a real number, $-a$, such that $a + (-a) = 0$
	6.	For every real number, $a \neq 0$, there exists a real number, $1/a$, such that $a\ (1/a) = 1$

An equation is a statement of equality. For example, $4 + 3 = 7$. An equation can also be written with one or more unknowns (or variables). The equation $x + 7 = 9$ is an equality only when the unknown $x = 2$. The number 2 is called the root or solution of this equation.

The end product of algebra is solving a mathematical equation(s). The operator normally will be involved in the solution of equations that are either linear, quadratic, or simultaneous in nature.

Summary

The important information in this chapter is summarized below.

Algebraic Laws Summary

Commutative Law of Addition	$a + b = b + a$
Commutative Law of Multiplication	$a(b) = b(a)$
Associative Law of Addition	$a+(b+c) = (a+b)+c$
Associative Law of Multiplication	$a(bc) = (ab)c$
Distributive Law	$a(b + c) = ab + ac$

LINEAR EQUATIONS

This chapter covers solving for unknowns using linear equations.

EO 1.2 SOLVE for the unknown given a linear equation.

The rules for addition, subtraction, multiplication, and division described in previous lessons will apply when solving linear equations. Before continuing this course it may be worthwhile to review the basic math laws in Module 1 and the first chapter of this module.

Solutions to Algebraic Equations

The equation is the most important concept in mathematics. Alone, algebraic operations are of little practical value. Only when these operations are coupled with algebraic equations can algebra be applied to solve practical problems.

An equation is a statement of equality between two equal quantities. Most people are familiar with the concept of equality. The idea of equal physical quantities is encountered routinely. An equation is merely the statement of this equality. There are three key ideas in an equation: an equation must involve two expressions, the expressions must be equal, and the equation must indicate that the expressions are equal. Thus, the statement that the sum of three and one equals four is an equation. It involves two expressions, (four and the sum of three and one), the expressions are equal, and the equation states that they are equal.

The equal sign (=) is used to indicate equality in an equation. In its most general form, an algebraic equation consists of two algebraic expressions separated by an equal sign. The equal sign is the key sign in algebra. It is the sign that defines one expression in terms of another. In solving practical problems, it is the sign that defines the unknown quantity in terms of known quantities.

Algebraic Equations

There are two kinds of equations: identities and conditional equations. An identity is an equation that is true for all values of the unknown involved. The identity sign (≡) is used in place of the equal sign to indicate an identity. Thus, $x^2 \equiv (x)(x)$, $3y + 5y \equiv 8y$, and $yx + yz \equiv y(x + z)$ are all identities because they are true for all values of x, y, or z. A conditional equation is one that is true only for some particular value(s) of the literal number(s) involved. A conditional equation is $3x + 5 = 8$, because only the value $x = 1$ satisfies the equation. When the word equation is used by itself, it usually means a conditional equation.

The root(s) of an equation (conditional equation) is any value(s) of the literal number(s) in the equation that makes the equation true. Thus, 1 is the root of the equation $3x + 5 = 8$ because $x = 1$ makes the equation true. To solve an algebraic equation means to find the root(s) of the equation.

The application of algebra is practical because many physical problems can be solved using algebraic equations. For example, pressure is defined as the force that is applied divided by the area over which it is applied. Using the literal numbers P (to represent the pressure), F (to represent the force), and A (to represent the area over which the force is applied), this physical relationship can be written as the algebraic equation $P = \frac{F}{A}$. When the numerical values of the force, F, and the area, A, are known at a particular time, the pressure, P, can be computed by solving this algebraic equation. Although this is a straightforward application of an algebraic equation to the solution of a physical problem, it illustrates the general approach that is used. Almost all physical problems are solved using this approach.

Types of Algebraic Equations

The letters in algebraic equations are referred to as unknowns. Thus, x is the unknown in the equation $3x + 5 = 8$. Algebraic equations can have any number of unknowns. The name unknown arises because letters are substituted for the numerical values that are not known in a problem.

The number of unknowns in a problem determines the number of equations needed to solve for the numerical values of the unknowns. Problems involving one unknown can be solved with one equation, problems involving two unknowns require two independent equations, and so on.

The degree of an equation depends on the power of the unknowns. The degree of an algebraic term is equivalent to the exponent of the unknown. Thus, the term $3x$ is a first degree term; $3x^2$ is a second degree term, and $3x^3$ is a third degree term. The degree of an equation is the same as the highest degree term. Linear or first degree equations contain no terms higher than first degree. Thus, $2x + 3 = 9$ is a linear equation. Quadratic or second degree equations contain up to second degree terms, but no higher. Thus, $x^2 + 3x = 6$, is a quadratic equation. Cubic or third degree equations contain up to third degree terms, but no higher. Thus, $4x^3 + 3x = 12$ is a cubic equation.

The degree of an equation determines the number of roots of the equation. Linear equations have one root, quadratic equations have two roots, and so on. In general, the number of roots of any equation is the same as the degree of the equation.

Exponential equations are those in which the unknown appears in the exponent. For example, $e^{-2.7x} = 290$ is an exponential equation. Exponential equations can be of any degree.

The basic principle used in solving any algebraic equation is: any operation performed on one side of an equation must also be performed on the other side for the equation to remain true. This one principle is used to solve all types of equations.

There are four axioms used in solving equations:

Axiom 1. If the same quantity is added to both sides of an equation, the resulting equation is still true.

Axiom 2. If the same quantity is subtracted from both sides of an equation, the resulting equation is still true.

Axiom 3. If both sides of an equation are multiplied by the same quantity, the resulting equation is still true.

Axiom 4. If both sides of an equation are divided by the same quantity, except 0, the resulting equation is still true.

Axiom 1 is called the addition axiom; Axiom 2, the subtraction axiom; Axiom 3, the multiplication axiom; and Axiom 4, the division axiom. These four axioms can be visualized by the balancing of a scale. If the scale is initially balanced, it will remain balanced if the same weight is added to both sides, if the same weight is removed from both sides, if the weights on both sides are increased by the same factor, or if the weights on both sides are decreased by the same factor.

Linear Equations

These four axioms are used to solve linear equations with three steps:

Step 1. Using the addition and subtraction axioms, Axioms 1 and 2, eliminate all terms with no unknowns from the left-hand side of the equation and eliminate all terms with the unknowns from the right-hand side of the equation.

Step 2. Using the multiplication and division axioms, Axioms 3 and 4, eliminate the coefficient from the unknowns on the left-hand side of the equation.

Step 3. Check the root by substituting it for the unknowns in the original equation.

Example 1:

Solve the equation $3x + 7 = 13$.

Solution:

Step 1. Using Axiom 2, subtract 7 from both sides of the equation.

$$3x + 7 - 7 = 13 - 7$$
$$3x = 6$$

Step 2. Using Axiom 4, divide both sides of the equation by 3.

$$\frac{3x}{3} = \frac{6}{3}$$

$$x = 2$$

Step 3. Check the root.

$$3(2) + 7 = 6 + 7 = 13$$

The root checks.

Example 2:

Solve the equation $2x + 9 = 3(x + 4)$.

Solution:

Step 1. Using Axiom 2, subtract $3x$ and 9 from both sides of the equation.

$$2x + 9 = 3(x + 4)$$
$$2x + 9 - 3x - 9 = 3x + 12 - 3x - 9$$
$$-x = 3$$

Step 2. Using Axiom 4, divide both sides of the equation by -1.

$$\frac{-x}{-1} = \frac{3}{-1}$$
$$x = -3$$

Step 3. Check the root.

$$2(-3) + 9 = -6 + 9 = 3$$
$$3[(-3) + 4] = 3(1) = 3$$

The root checks.

These same steps can be used to solve equations that include several unknowns. The result is an expression for one of the unknowns in terms of the other unknowns. This is particularly important in solving practical problems. Often the known relationship among several physical quantities must be rearranged in order to solve for the unknown quantity. The steps are performed so that the unknown quantity is isolated on the left-hand side of the equation.

Example 1:

Solve the equation $ax - b = c$ for x in terms of a, b, and c.

Solution:

Step 1. Using Axiom 1, add b to both sides of the equation.

$$ax - b + b = c + b$$
$$ax = c + b$$

Step 2. Using Axiom 4, divide both sides of the equation by a.

$$\frac{ax}{a} = \frac{c + b}{a}$$

$$x = \frac{c + b}{a}$$

Step 3. Check the root.

$$a\frac{c + b}{a} - b = c + b - b = c$$

The root checks.

Example 2:

The equation relating the pressure, P, to the force, F, and the area, A, over which the force is applied is $P = \frac{F}{A}$. Solve this equation for F, in terms of P and A.

Solution:

Step 1. Axioms 1 and 2 do not help solve the problem, so go to Step 2.

Step 2. Using Axiom 3, multiply both sides of the equation by A.

$$P(A) = \frac{F}{A}(A)$$

$$F = PA$$

Step 3. Check the root.

$$\frac{PA}{A} = P$$

The root checks.

The addition or subtraction of the same quantity from both sides of an equation may be accomplished by transposing a quantity from one side of the equation to the other. Transposing is a shortened way of applying the addition or subtraction axioms. Any term may be transposed or transferred from one side of an equation to the other if its sign is changed. Thus, in the equation $5x + 4 = 7$, the 4 can be transposed to the other side of the equation by changing its sign. The result is $5x = 7 - 4$ or $5x = 3$. This corresponds to applying the subtraction axiom, Axiom 2, subtracting 4 from both sides of the equation.

Example:

Solve the equation $4x + 3 = 19$ by transposing.

Solution:

Step 1. Transpose the 3 from the left-hand to the right-hand side of the equation by changing its sign.

$$4x = 19 - 3$$

$$4x = 16$$

Step 2. Using Axiom 4, divide both sides of the equation by 4.

$$\frac{4x}{4} = \frac{16}{4}$$

$$x = 4$$

Step 3. Check the root.

$$4(4) + 3 = 16 + 3 = 19$$

The root checks.

Solving Fractional Equations

A fractional equation is an equation containing a fraction. The fraction can be either a common fraction or a decimal fraction. The unknowns can occupy any position in the equation. They may or may not be part of the fraction. If they are part of the fraction, they can be either in the numerator or the denominator. The following are three examples of fractional equations:

$$5x + \frac{1}{2} = 8 \qquad \frac{2x + 6}{3x} = 9 - y \qquad 0.67x + 1.25y = 9$$

Fractional equations are solved using the same axioms and approach used for other algebraic equations. However, the initial step is to remove the equation from fractional form. This is done by determining the lowest common denominator (LCD) for all of the fractions in the equation and then multiplying both sides of the equation by this common denominator. This will clear the equation of fractions.

Example 1:

Solve the fractional equation $\frac{3x + 8}{x} + 5 = 0$.

Solution:

Multiply both sides of the equation by the LCD (x).

$$(x)\left(\frac{3x + 8}{x} + 5\right) = (0)(x)$$

$$3x + 8 + 5x = 0$$

$$8x + 8 = 0$$

Now solve the equation like an ordinary linear equation.

Step 1. Transpose the +8 from the left-hand to the right-hand side of the equation by changing its sign.

$$8x = 0 - 8$$

$$8x = -8$$

Step 2. Using Axiom 4, divide both sides of the equation by 8.

$$\frac{8x}{8} = \frac{-8}{8}$$

$$x = -1$$

Step 3. Check the root.

$$\frac{3(-1) + 8}{-1} + 5 = \frac{-3 + 8}{-1} + 5 = -5 + 5 = 0$$

The root checks.

Example 2:

Solve the fractional equation $\frac{1}{x-2} + \frac{1}{x+3} = 0$

Solution:

The LCD is $(x - 2)(x + 3)$; therefore, multiply both sides of the equation by $(x - 2)(x + 3)$.

$$(x-2)(x+3)\left(\frac{1}{x-2} + \frac{1}{x+3}\right) = (0)(x-2)(x+3)$$

$$\frac{(x-2)(x+3)}{(x-2)} + \frac{(x-2)(x+3)}{(x+3)} = 0$$

$$(x + 3) + (x - 2) = 0$$

$$2x + 1 = 0$$

Now solve the equation like an ordinary linear equation.

Step 1. Transpose the +1 from the left-hand to the right-hand side of the equation by changing its sign.

$$2x = 0 - 1$$

$$2x = - 1$$

Step 2. Using Axiom 4, divide both sides of the equation by 2.

$$\frac{2x}{2} = \frac{-1}{2}$$

$$x = -\frac{1}{2}$$

Step 3. Check the root.

$$\frac{1}{-\frac{1}{2}-2}+\frac{1}{-\frac{1}{2}+3}=\frac{1}{-2\frac{1}{2}}+\frac{1}{2\frac{1}{2}}=-\frac{2}{5}+\frac{2}{5}=0$$

The root checks.

Ratio and Proportion

One of the most important applications of fractional equations is ratio and proportion. A ratio is a comparison of two like quantities by division. It is written by separating the quantities by a colon or by writing them as a fraction. To write a ratio, the two quantities compared must be of the same kind. For example, the ratio of \$8 to \$12 is written as \$8:\$12 or $\frac{\$8}{\$12}$. Two unlike quantities cannot be compared by a ratio. For example, 1 inch and 30 minutes cannot form a ratio. However, two different units can be compared by a ratio if they measure the same kind of quantity. For example, 1 minute and 30 seconds can form a ratio, but they must first be converted to the same units. Since 1 minute equals 60 seconds, the ratio of 1 minute to 30 seconds is written 60 seconds:30 seconds, or $\frac{60 \text{ seconds}}{30 \text{ seconds}}$, which equals 2:1 or 2.

A proportion is a statement of equality between two ratios. For example, if a car travels 40 miles in 1 hour and 80 miles in 2 hours, the ratio of the distance traveled is 40 miles:80 miles, or $\frac{40 \text{ miles}}{80 \text{ miles}}$, and the ratio of time is 1 hour:2 hours, or $\frac{1 \text{ hour}}{2 \text{ hours}}$. The proportion relating these two ratios is:

40 miles:80 miles = 1 hour:2 hours

$$\frac{40 \text{ miles}}{80 \text{ miles}}=\frac{1 \text{ hour}}{2 \text{ hours}}$$

A proportion consists of four terms. The first and fourth terms are called the extremes of the proportion; the second and third terms are called the means. If the letters *a*, *b*, *c* and *d* are used to represent the terms in a proportion, it can be written in general form.

$$\frac{a}{b}=\frac{c}{d}$$

Multiplication of both sides of this equation by *bd* results in the following.

$$(bd)\frac{a}{b} = \frac{c}{d}(bd)$$

$$ad = cb$$

Thus, the product of the extremes of a proportion (*ad*) equals the product of the means (*bc*). For example, in the proportion 40 miles:80 miles = 1 hour:2 hours, the product of the extremes is (40 miles)(2 hours) which equals 80 miles-hours, and the product of the means is (80 miles)(1 hour), which also equals 80 miles-hours.

Ratio and proportion are familiar ideas. Many people use them without realizing it. When a recipe calls for 1½ cups of flour to make a serving for 6 people, and the cook wants to determine how many cups of flour to use to make a serving for 8 people, she uses the concepts of ratios and proportions. When the price of onions is 2 pounds for 49 cents and the cost of 3½ pounds is computed, ratio and proportion are used. Most people know how to solve ratio and proportion problems such as these without knowing the specific steps used.

Ratio and proportion problems are solved by using an unknown such as x for the missing term. The resulting proportion is solved for the value of x by setting the product of the extremes equal to the product of the means.

Example 1:

Solve the following proportion for x.

Solution:

$$5:x = 4:15$$

The product of the extremes is (5)(15) = 75.
The product of the means is $(x)(4) = 4x$.
Equate these two products and solve the resulting equation.

$$4x = 75$$

$$\frac{4x}{4} = \frac{75}{4}$$

$$x = 18\frac{3}{4}$$

Example 2:

If 5 pounds of apples cost 80 cents, how much will 7 pounds cost?

Solution:

Using x for the cost of 7 pounds of apples, the following proportion can be written.

$$\frac{5 \text{ pounds}}{7 \text{ pounds}} = \frac{80 \text{ cents}}{x}$$

The product of the extremes is $(5)(x) = 5x$.
The product of the means is $(7)(80) = 560$.
Equate these two products and solve the resulting equation.

$$5x = 560$$

$$\frac{5x}{5} = \frac{560}{5}$$

$$x = 112$$

The unit of x is cents. Thus, 7 pounds of apples cost 112 cents or \$1.12.

Example 3:

A recipe calls for $1\frac{1}{2}$ cups of flour to make servings for 6 people. How much flour should be used to make servings for 4 people?

Solution:

Using x for the flour required for 4 people, the following proportion can be written.

$$\frac{6 \text{ people}}{4 \text{ people}} = \frac{1\frac{1}{2} \text{ cups}}{x}$$

The product of the extremes is $(6)(x) = 6x$.

The product of the means is $(4)\ 1\frac{1}{2} = 6$.

Equate these two products and solve the resulting equation.

$$6x = 6$$

$$\frac{6x}{6} = \frac{6}{6}$$

$$x = 1$$

The unit of x is cups. Thus, servings for 4 people require 1 cup of flour.

Summary

The important information in this chapter is summarized below.

Linear Equations Summary

There are four axioms used in solving linear equations.

Axiom 1. If the same quantity is added to both sides of an equation, the resulting equation is still true.

Axiom 2. If the same quantity is subtracted from both sides of an equation, the resulting equation is still true.

Axiom 3. If both sides of an equation are multiplied by the same quantity, the resulting equation is still true.

Axiom 4. If both sides of an equation are divided by the same quantity, except 0, the resulting equation is still true.

Axiom 1 is called the addition axiom; Axiom 2, the subtraction axiom; Axiom 3, the multiplication axiom; and Axiom 4, the division axiom.

QUADRATIC EQUATIONS

This chapter covers solving for unknowns using quadratic equations.

EO 1.3 APPLY the quadratic formula to solve for an unknown.

Types of Quadratic Equations

A quadratic equation is an equation containing the second power of an unknown but no higher power. The equation $x^2 - 5x + 6 = 0$ is a quadratic equation. A quadratic equation has two roots, both of which satisfy the equation. The two roots of the quadratic equation $x^2 - 5x + 6 = 0$ are $x = 2$ and $x = 3$. Substituting either of these values for x in the equation makes it true.

The general form of a quadratic equation is the following:

$$ax^2 - bx + c = 0 \qquad (2\text{-}1)$$

The a represents the numerical coefficient of x^2 , b represents the numerical coefficient of x, and c represents the constant numerical term. One or both of the last two numerical coefficients may be zero. The numerical coefficient a cannot be zero. If b=0, then the quadratic equation is termed a "pure" quadratic equation. If the equation contains both an x and x^2 term, then it is a "complete" quadratic equation. The numerical coefficient c may or may not be zero in a complete quadratic equation. Thus, $x^2 + 5x + 6 = 0$ and $2x^2 - 5x = 0$ are complete quadratic equations.

Solving Quadratic Equations

The four axioms used in solving linear equations are also used in solving quadratic equations. However, there are certain additional rules used when solving quadratic equations. There are three different techniques used for solving quadratic equations: taking the square root, factoring, and the Quadratic Formula. Of these three techniques, only the Quadratic Formula will solve all quadratic equations. The other two techniques can be used only in certain cases. To determine which technique can be used, the equation must be written in general form:

$$ax^2 + bx + c = 0 \qquad (2\text{-}1)$$

If the equation is a pure quadratic equation, it can be solved by taking the square root. If the numerical constant c is zero, equation 2-1 can be solved by factoring. Certain other equations can also be solved by factoring.

Taking Square Root

A pure quadratic equation can be solved by taking the square root of both sides of the equation. Before taking the square root, the equation must be arranged with the x^2 term isolated on the left-hand side of the equation and its coefficient reduced to 1. There are four steps in solving pure quadratic equations by taking the square root.

Step 1. Using the addition and subtraction axioms, isolate the x^2 term on the left-hand side of the equation.

Step 2. Using the multiplication and division axioms, eliminate the coefficient from the x^2 term.

Step 3. Take the square root of both sides of the equation.

Step 4. Check the roots.

In taking the square root of both sides of the equation, there are two values that satisfy the equation. For example, the square roots of x^2 are $+x$ and $-x$ since $(+x)(+x) = x^2$ and $(-x)(-x) = x^2$. The square roots of 25 are +5 and -5 since $(+5)(+5) = 25$ and $(-5)(-5) = 25$. The two square roots are sometimes indicated by the symbol ±. Thus, $\sqrt{25} = \pm 5$. Because of this property of square roots, the two roots of a pure quadratic equation are the same except for their sign.

At this point, it should be mentioned that in some cases the result of solving pure quadratic equations is the square root of a negative number. Square roots of negative numbers are called imaginary numbers and will be discussed later in this section.

Example:

Solve the following quadratic equation by taking the square roots of both sides.

$$3x^2 = 100 - x^2$$

Solution:

Step 1. Using the addition axiom, add x^2 to both sides of the equation.

$$3x^2 + x^2 = 100 - x^2 + x^2$$

$$4x^2 = 100$$

Step 2. Using the division axiom, divide both sides of the equation by 4.

$$\frac{4x^2}{4} = \frac{100}{4}$$

$$x^2 = 25$$

Step 3. Take the square root of both sides of the equation.

$$x^2 = 25$$

$$\sqrt{x^2} = \sqrt{25}$$

$$x = \pm 5$$

Thus, the roots are $x = +5$ and $x = -5$.

Step 4. Check the roots.

$$3x^2 = 100 - x^2$$

$$3(\pm 5)^2 = 100 - (\pm 5)^2$$

$$3(25) = 100 - 25$$

$$75 = 75$$

If a pure quadratic equation is written in general form, a general expression can be written for its roots. The general form of a pure quadratic is the following.

$$ax^2 + c = 0 \qquad (2\text{-}2)$$

Using the subtraction axiom, subtract c from both sides of the equation.

$$ax^2 = -c$$

Using the division axiom, divide both sides of the equation by a.

$$x^2 = -\frac{c}{a}$$

Now take the square roots of both sides of the equation.

$$x = +\sqrt{-\frac{c}{a}} \qquad (2\text{-}3)$$

Thus, the roots of a pure quadratic equation written in general form $ax^2 + c = 0$ are $x = +\sqrt{-\frac{c}{a}}$ and $x = -\sqrt{-\frac{c}{a}}$.

Example:

Find the roots of the following pure quadratic equation.

$$4x^2 - 100 = 0$$

Solution:

Using Equation 2-3, substitute the values of c and a and solve for x.

$$x = +\sqrt{-\frac{c}{a}}$$

$$x = +\sqrt{-\frac{(-100)}{4}}$$

$$x = \pm\sqrt{25}$$

$$x = \pm 5$$

Thus, the roots are $x = 5$ and $x = -5$.

Factoring Quadratic Equations

Certain complete quadratic equations can be solved by factoring. If the left-hand side of the general form of a quadratic equation can be factored, the only way for the factored equation to be true is for one or both of the factors to be zero. For example, the left-hand side of the quadratic equation $x^2 + x - 6 = 0$ can be factored into $(x + 3)(x - 2)$. The only way for the equation $(x + 3)(x - 2) = 0$ to be true is for either $(x + 3)$ or $(x - 2)$ to be zero. Thus, the roots of quadratic equations which can be factored can be found by setting each of the factors equal to zero and solving the resulting linear equations. Thus, the roots of $(x + 3)(x - 2) = 0$ are found by setting $x + 3$ and $x - 2$ equal to zero. The roots are $x = -3$ and $x = 2$.

Factoring estimates can be made on the basis that it is the reverse of multiplication. For example, if we have two expressions $(dx + c)$ and $(cx + g)$ and multiply them, we obtain (using the distribution laws)

$$(dx + c)(fx + g) = (dx)(fx) + (dx)(g) + (c)(fx) + cg =$$
$$= dfx^2 + (dg + cf)x + cg.$$

Thus, a statement $(dx + c)(fx + g) = 0$ can be written

$$df\,x^2 + (dg + cf)x + cg = 0.$$

Now, if one is given an equation $ax^2 + bx + c = 0$, he knows that the symbol a is the product of two numbers (df) and c is also the product of two numbers. For the example $3x^2 - 4x - 4 = 0$, it is a reasonable guess that the numbers multiplying x^2 in the two factors are 3 and 1, although they might be 1.5 and 2. The last -4 (c in the general equation) is the product of two numbers (eg), perhaps -2 and 2 or -1 and 4. These combinations are tried to see which gives the proper value of b $(dg + ef)$, from above.

There are four steps used in solving quadratic equations by factoring.

Step 1. Using the addition and subtraction axioms, arrange the equation in the general quadratic form $ax^2 + bx + c = 0$.

Step 2. Factor the left-hand side of the equation.

Step 3. Set each factor equal to zero and solve the resulting linear equations.

Step 4. Check the roots.

Example:

Solve the following quadratic equation by factoring.

$$2x^2 - 3 = 4x - x^2 + 1$$

Solution:

Step 1. Using the subtraction axiom, subtract $(4x - x^2 + 1)$ from both sides of the equation.

$$2x^2 - 3 - (4x - x^2 + 1) = 4x - x^2 + 1 - (4x - x^2 + 1)$$

$$3x^2 - 4x - 4 = 0$$

Step 2. Factor the resulting equation.

$$3x^2 - 4x - 4 = 0$$

$$(3x + 2)(x - 2) = 0$$

Step 3. Set each factor equal to zero and solve the resulting equations.

$$3x + 2 = 0$$

$$3x = -2$$

$$\frac{3x}{3} = \frac{-2}{3}$$

$$x = -\frac{2}{3}$$

$$x - 2 = 0$$

$$x = 2$$

Thus, the roots are $x = -\frac{2}{3}$ and $x = 2$.

Step 4. Check the roots.

$$2x^2 - 3 = 4x - x^2 + 1$$

$$2\left(-\frac{2}{3}\right)^2 - 3 = 4\left(-\frac{2}{3}\right) - \left(-\frac{2}{3}\right)^2 + 1$$

$$2\left(\frac{4}{9}\right) - 3 = -\frac{8}{3} - \frac{4}{9} + 1$$

$$\frac{8}{9} - \frac{27}{9} = -\frac{24}{9} - \frac{4}{9} + \frac{9}{9}$$

$$-\frac{19}{9} = -\frac{19}{9}$$

$$2x^2 - 3 = 4x - x^2 + 1$$

$$2(2)^2 - 3 = 4(2) - (2)^2 + 1$$

$$2(4) - 3 = 8 - 4 + 1$$

$$8 - 3 = 5$$

$$5 = 5$$

Thus, the roots check.

Quadratic equations in which the numerical constant c is zero can always be solved by factoring. One of the two roots is zero. For example, the quadratic equation $2x^2 + 3x = 0$ can be solved by factoring. The factors are (x) and $(2x + 3)$. Thus, the roots are $x = 0$ and $x = -\frac{3}{2}$. If a quadratic equation in which the numerical constant c is zero is written in general form, a general expression can be written for its roots. The general form of a quadratic equation in which the numerical constant c is zero is the following:

$$ax^2 + bx = 0 \qquad (2\text{-}4)$$

The left-hand side of this equation can be factored by removing an x from each term.

$$x(ax + b) = 0 \qquad (2\text{-}5)$$

The roots of this quadratic equation are found by setting the two factors equal to zero and solving the resulting equations.

$$x = 0 \tag{2-6}$$

$$x = -\frac{b}{a} \tag{2-7}$$

Thus, the roots of a quadratic equation in which the numerical constant c is zero are $x = 0$ and $x = -\frac{b}{a}$.

Example:

Find the roots of the following quadratic equation.

$$3x^2 + 7x = 0$$

Solution:

Using Equation 2-6, one root is determined.

$$x = 0$$

Using Equation 2-7, substitute the values of a and b and solve for x.

$$x = -\frac{b}{a}$$

$$x = -\frac{7}{3}$$

Thus, the roots are $x = 0$ and $x = -\frac{7}{3}$.

The Quadratic Formula

Many quadratic equations cannot readily be solved by either of the two techniques already described (taking the square roots or factoring). For example, the quadratic equation $x^2 - 6x + 4 = 0$ is not a pure quadratic and, therefore, cannot be solved by taking the square roots. In addition, the left-hand side of the equation cannot readily be factored. The Quadratic Formula is a third technique for solving quadratic equations. It can be used to find the roots of any quadratic equation.

$$x = \frac{-b \pm \sqrt{b^2 - 4ac}}{2a} \tag{2-8}$$

Equation 2-8 is the Quadratic Formula. It states that the two roots of a quadratic equation written in general form, $ax^2 + bx + c = 0$, are equal to $x = \frac{-b + \sqrt{b^2 - 4ac}}{2a}$ and $x = \frac{-b - \sqrt{b^2 - 4ac}}{2a}$. The Quadratic Formula should be committed to memory because it is such a useful tool for solving quadratic equations.

There are three steps in solving a quadratic equation using the Quadratic Formula.

Step 1. Write the equation in general form.

Step 2. Substitute the values for a, b, and c into the Quadratic Formula and solve for x.

Step 3. Check the roots in the original equation.

Example 1:

Solve the following quadratic equation using the Quadratic Formula.

$4x^2 + 2 = x^2 - 7x$:

Solution:

Step 1. Write the equation in general form.

$$4x^2 + 2 = x^2 - 7x$$

$$3x^2 + 7x + 2 = 0$$

$$a = +3,\ b = +7,\ c = +2$$

$$x = \frac{-b \pm \sqrt{b^2 - 4ac}}{2a}$$

$$x = \frac{-7 \pm \sqrt{(7)^2 - 4(3)(2)}}{2(3)}$$

$$x = \frac{-7 \pm \sqrt{49 - 24}}{6}$$

Step 2.

$$x = \frac{-7 \pm \sqrt{25}}{6}$$

$$x = \frac{-7 \pm 5}{6}$$

$$x = \frac{-7 + 5}{6},\ \frac{-7 - 5}{6}$$

$$x = \frac{-2}{6},\ \frac{-12}{6}$$

$$x = -\frac{1}{3},\ -2$$

Thus, the roots are $x = -\frac{1}{3}$ and $x = -2$.

Step 3. Check the roots.

$$4x^2 + 2 = x^2 - 7x$$

$$4\left(-\frac{1}{3}\right)^2 + 2 = \left(-\frac{1}{3}\right)^2 - 7\left(-\frac{1}{3}\right)$$

$$4\left(\frac{1}{9}\right) + 2 = \frac{1}{9} - \left(-\frac{7}{3}\right)$$

$$\frac{4}{9} + \frac{18}{9} = \frac{1}{9} + \frac{21}{9}$$

$$\frac{22}{9} = \frac{22}{9}$$

and,

$$4x^2 + 2 = x^2 - 7x$$

$$4(-2)^2 + 2 = (-2)^2 - 7(-2)$$

$$4(4) + 2 = 4 -(-14)$$

$$16 + 2 = 4 + 14$$

$$18 = 18$$

Thus, the roots check.

Example 2:

Solve the following quadratic equation using the Quadratic Formula.

$$2x^2 + 4 = 6x + x^2$$

Solution:

Step 1. Write the equation in general form.

$$2x^2 + 4 = 6x + x^2$$

$$x^2 - 6x + 4 = 0$$

$$a = +1,\ b = -6,\ c = +4$$

$$x = \frac{-b \pm \sqrt{b^2 - 4ac}}{2a}$$

$$x = \frac{-(-6) \pm \sqrt{(-6)^2 - 4(1)(4)}}{2(1)}$$

$$x = \frac{6 \pm \sqrt{36 - 16}}{2}$$

$$x = \frac{6 \pm \sqrt{20}}{2}$$

Step 2.

$$x = 3 \pm \frac{1}{2}\sqrt{20}$$

$$x = 3 \pm \frac{1}{2}\sqrt{(4)(5)}$$

$$x = 3 \pm \sqrt{5}$$

$$x = 3 + \sqrt{5},\ 3 - \sqrt{5}$$

$$x = 3 + 2.236,\ 3 - 2.236$$

$$x = 5.236,\ 0.746$$

Step 3. Check the roots.

$$2x^2 + 4 = 6x + x^2$$

$$2(3 + \sqrt{5})^2 + 4 = 6(3 + \sqrt{5}) + (3 + \sqrt{5})^2$$

$$2(9 + 6\sqrt{5} + 5) + 4 = 18 + 6\sqrt{5} + 9 + 6\sqrt{5} + 5$$

$$18 + 12\sqrt{5} + 10 + 4 = 18 + 12\sqrt{5} + 9 + 5$$

$$32 + 12\sqrt{5} = 32 + 12\sqrt{5}$$

and,

$$2x^2 + 4 = 6x + x^2$$

$$2(3 - \sqrt{5})^2 + 4 = 6(3 - \sqrt{5}) + (3 - \sqrt{5})^2$$

$$2(9 - 6\sqrt{5} + 5) + 4 = 18 - 6\sqrt{5} + 9 - 6\sqrt{5} + 5$$

$$18 - 12\sqrt{5} + 10 + 4 = 18 - 12\sqrt{5} + 9 + 5$$

$$32 - 12\sqrt{5} = 32 - 12\sqrt{5}$$

Thus, the roots check.

The Quadratic Formula can be used to find the roots of any quadratic equation. For a pure quadratic equation in which the numerical coefficient b equals zero, the Quadratic Formula (2-8) reduces to the formula given as Equation 2-9.

$$x = \frac{-b \pm \sqrt{b^2 - 4ac}}{2a} \qquad (2\text{-}8)$$

For $b = 0$, this reduces to the following.

$$x = \frac{\pm\sqrt{-4ac}}{2a}$$

$$x = \pm\sqrt{\frac{-4ac}{4a^2}}$$

$$x = \pm\sqrt{-\frac{c}{a}} \quad (2\text{-}9)$$

Summary

The important information in this chapter is summarized below.

Quadratic Equations Summary

There are three methods used when solving quadratic equations:

- Taking the square root
- Factoring the equation
- Using the quadratic formula

$$x = \frac{-b\pm\sqrt{b^2-4ac}}{2a}$$

SIMULTANEOUS EQUATIONS

This chapter covers solving for two unknowns using simultaneous equations.

EO 1.4 Given simultaneous equations, SOLVE for the unknowns.

Many practical problems that can be solved using algebraic equations involve more than one unknown quantity. These problems require writing and solving several equations, each of which contains one or more of the unknown quantities. The equations that result in such problems are called simultaneous equations because all the equations must be solved simultaneously in order to determine the value of any of the unknowns. The group of equations used to solve such problems is called a system of equations.

The number of equations required to solve any problem usually equals the number of unknown quantities. Thus, if a problem involves only one unknown, it can be solved with a single equation. If a problem involves two unknowns, two equations are required. The equation $x + 3 = 8$ is an equation containing one unknown. It is true for only one value of x: $x = 5$. The equation $x + y = 8$ is an equation containing two unknowns. It is true for an infinite set of xs and ys. For example: $x = 1, y = 7$; $x = 2, y = 6$; $x = 3, y = 5$; and $x = 4, y = 4$ are just a few of the possible solutions. For a system of two linear equations each containing the same two unknowns, there is a single pair of numbers, called the solution to the system of equations, that satisfies both equations. The following is a system of two linear equations:

$$2x + y = 9$$
$$x - y = 3$$

The solution to this system of equations is $x = 4, y = 1$ because these values of x and y satisfy both equations. Other combinations may satisfy one or the other, but only $x = 4, y = 1$ satisfies both.

Systems of equations are solved using the same four axioms used to solve a single algebraic equation. However, there are several important extensions of these axioms that apply to systems of equations. These four axioms deal with adding, subtracting, multiplying, and dividing both sides of an equation by the same quantity. The left-hand side and the right-hand side of any equation are equal. They constitute the same quantity, but are expressed differently. Thus, the left-hand and right-hand sides of one equation can be added to, subtracted from, or used to multiply or divide the left-hand and right-hand sides of another equation, and the resulting equation will still be true. For example, two equations can be added.

$$\begin{array}{rcrcr} 3x & + & 4y & = & 7 \\ +\ (x & + & 5y & = & 12) \\ \hline 4x & + & 9y & = & 19 \end{array}$$

Adding the second equation to the first corresponds to adding the same quantity to both sides of the first equation. Thus, the resulting equation is still true. Similarly, two equations can be subtracted.

$$\begin{array}{rcrcr} 4x & - & 3y & = & 8 \\ -\ (2x & + & 5y & = & 11) \\ \hline 2x & - & 8y & = & -3 \end{array}$$

Subtracting the second equation from the first corresponds to subtracting the same quantity from both sides of the first equation. Thus, the resulting equation is still true.

The basic approach used to solve a system of equations is to reduce the system by eliminating the unknowns one at a time until one equation with one unknown results. This equation is solved and its value used to determine the values of the other unknowns, again one at a time. There are three different techniques used to eliminate unknowns in systems of equations: addition or subtraction, substitution, and comparison.

Solving Simultaneous Equations

The simplest system of equations is one involving two linear equations with two unknowns.

$$\begin{aligned} 5x + 6y &= 12 \\ 3x + 5y &= 3 \end{aligned}$$

The approach used to solve systems of two linear equations involving two unknowns is to combine the two equations in such a way that one of the unknowns is eliminated. The resulting equation can be solved for one unknown, and either of the original equations can then be used to solve for the other unknown.

Systems of two equations involving two unknowns can be solved by addition or subtraction using five steps.

Step 1. Multiply or divide one or both equations by some factor or factors that will make the coefficients of one unknown numerically equal in both equations.

Step 2. Eliminate the unknown having equal coefficients by addition or subtraction.

Step 3. Solve the resulting equation for the value of the one remaining unknown.

Step 4. Find the value of the other unknown by substituting the value of the first unknown into one of the original equations.

Step 5. Check the solution by substituting the values of the two unknowns into the other original equation.

Example:

Solve the following system of equations using addition or subtraction.

$$5x + 6y = 12$$

$$3x + 5y = 3$$

Solution:

Step 1. Make the coefficients of y equal in both equations by multiplying the first equation by 5 and the second equation by 6.

$$5(5x + 6y = 12) \text{ yields } 25x + 30y = 60$$

$$6(3x + 5y = 3) \text{ yields } 18x + 30y = 18$$

Step 2. Subtract the second equation from the first.

$$\begin{array}{r} 25x + 30y = 60 \\ -(18x + 30y = 18) \\ \hline 7x + 0 = 42 \end{array}$$

Step 3. Solve the resulting equation.

$$7x = 42$$

$$\frac{7x}{7} = \frac{42}{7}$$

$$x = 6$$

Step 4. Substitute $x = 6$ into one of the original equations and solve for y.

$$5x + 6y = 12$$

$$5(6) + 6y = 12$$

$$30 + 6y = 12$$

$$6y = 12 - 30$$

$$6y = -18$$

$$\frac{6y}{6} = \frac{-18}{6}$$

$$y = -3$$

Step 5. Check the solution by substituting $x = 6$ and $y = -3$ into the other original equation.

$$3x + 5y = 3$$

$$3(6) + 5(-3) = 3$$

$$18 - 15 = 3$$

$$3 = 3$$

Thus, the solution checks.

Systems of two equations involving two unknowns can also be solved by substitution.

Step 1. Solve one equation for one unknown in terms of the other.

Step 2. Substitute this value into the other equation.

Step 3. Solve the resulting equation for the value of the one remaining unknown.

Step 4. Find the value of the other unknown by substituting the value of the first unknown into one of the original equations.

Step 5. Check the solution by substituting the values of the two unknowns into the other original equation.

Example:

Solve the following system of equations using substitution.

$$5x + 6y = 12$$
$$3x + 5y = 3$$

Solution:

Step 1. Solve the first equation for x.

$$5x + 6y = 12$$

$$5x = 12 - 6y$$

$$\frac{5x}{5} = \frac{12 - 6y}{5}$$

$$x = \frac{12}{5} - \frac{6y}{5}$$

Step 2. Substitute this value of x into the second equation.

$$3x + 5y = 3$$

$$3\left(\frac{12}{5} - \frac{6y}{5}\right) + 5y = 3$$

Step 3. Solve the resulting equation.

$$3\left(\frac{12}{5} - \frac{6y}{5}\right) + 5y = 3$$

$$\frac{36}{5} - \frac{18y}{5} + 5y = 3$$

$$(5)\left(\frac{36}{5} - \frac{18}{5}y + 5y\right) = 3(5)$$

$$36 - 18y + 25y = 15$$

$$7y = 15 - 36$$

$$7y = -21$$

$$\frac{7y}{7} = \frac{-21}{7}$$

$$y = -3$$

Step 4. Substitute $y = -3$ into one of the original equations and solve for x.

$$5x + 6y = 12$$

$$5x + 6(-3) = 12$$

$$5x - 18 = 12$$

$$5x = 12 + 18$$

$$5x = 30$$

$$\frac{5x}{5} = \frac{30}{5}$$

$$x = 6$$

Step 5. Check the solution by substituting $x = 6$ and $y = -3$ into the other original equation.

$$3x + 5y = 3$$
$$3(6) + 5(-3) = 3$$
$$18 - 15 = 3$$
$$3 = 3$$

Thus, the solution checks.

Systems of two equations involving two unknowns can also be solved by comparison.

Step 1. Solve each equation for the same unknown in terms of the other unknown.

Step 2. Set the two expressions obtained equal to each other.

Step 3. Solve the resulting equation for the one remaining unknown.

Step 4. Find the value of the other unknown by substituting the value of the first unknown into one of the original equations.

Step 5. Check the solution by substituting the values of the two unknowns into the other original equation.

Example:

Solve the following system of equations by comparison.

$$5x + 6y = 12$$
$$3x + 5y = 3$$

Solution:

Step 1. Solve both equations for x.

$$5x + 6y = 12$$

$$5x = 12 - 6y$$

$$\frac{5x}{5} = \frac{12 - 6y}{5}$$

$$x = \frac{12 - 6y}{5}$$

$$3x + 5y = 3$$

$$3x = 3 - 5y$$

$$\frac{3x}{3} = \frac{3 - 5y}{3}$$

$$x = \frac{3 - 5y}{3}$$

Step 2. Set the two values for x equal to each other.

$$\frac{12 - 6y}{5} = \frac{3 - 5y}{3}$$

Step 3. Solve the resulting equation for y.

$$\frac{12 - 6y}{5} = \frac{3 - 5y}{3}$$

$$(3)(5)\,\frac{12 - 6y}{5} = \frac{3 - 5y}{3}\,(3)(5)$$

$$3(12 - 6y) = 5(3 - 5y)$$

$$36 - 18y = 15 - 25y$$

$$25y - 18y = 15 - 36$$

$$7y = -21$$

$$\frac{7y}{7} = \frac{-21}{7}$$

$$y = -3$$

Step 4. Substitute $y = -3$ into one of the original equations and solve for x.

$$5x + 6y = 12$$

$$5x + 6(-3) = 12$$

$$5x - 18 = 12$$

$$5x = 12 + 18$$

$$5x = 30$$

$$\frac{5x}{5} = \frac{30}{5}$$

$$x = 6$$

Step 5. Check the solution by substituting $x = 6$ and $y = -3$ into the other original equation.

$$3x + 5y = 3$$

$$3(6) + 5(-3) = 3$$

$$18 - 15 = 3$$

$$3 = 3$$

Thus, the solution checks.

Quite often, when more than one unknown exists in a problem, the end result of the equations expressing the problem is a set of simultaneous equations showing the relationship of one of the unknowns to the other unknowns.

Example:

Solve the following simultaneous equations by substitution.

$$3x + 4y = 6 \qquad 5x + 3y = -1$$

Solution:

Solve for x:

$$3x = 6 - 4y$$

$$x = 2 - \frac{4y}{3}$$

Substitute the value for x into the other equation:

$$5\left(2 - \frac{4y}{3}\right) + 3y = -1$$

$$10 - \frac{20y}{3} + 3y = -1$$

$$10 - \frac{20y}{3} + \frac{9y}{3} = -1$$

$$10 - \frac{11y}{3} = -1$$

$$\frac{-11y}{3} = -11$$

$$y = 3$$

Substitute $y = 3$ into the first equation:

$$3x + 4(3) = 6$$

$$3x = -6$$

$$x = -2$$

Check the solution by substituting $x = -2$ and $y = 3$ into the original equations.

$3x + 4y = 6$	$5x + 3y = -1$
$3(-2) + 4(3) = 6$	$5(-2) + 3(3) = -1$
$-6 + 12 = 6$	$-10 + 9 = -1$
$6 = 6$	$-1 = -1$

Thus, the solution checks.

Summary

The important information in this chapter is summarized below.

Simultaneous Equations Summary

There are three methods used when solving simultaneous equations:

- Addition or subtraction
- Substitution
- Comparison

WORD PROBLEMS

This chapter covers ways of setting up word problems and solving for the unknowns.

EO 1.5 Given a word problem, write equations and SOLVE for the unknown.

Basic Approach to Solving Algebraic Word Problems

Algebra is used to solve problems in science, industry, business, and the home. Algebraic equations can be used to describe laws of motion, pressures of gases, electric circuits, and nuclear facility operations. They can be applied to problems about the ages of people, the cost of articles, football scores, and other everyday matters. The basic approach to solving problems in these apparently dissimilar fields is the same. First, condense the available information into algebraic equations, and, second, solve the equations. Of these two basic steps, the first is frequently the most difficult to master because there are no clearly defined rules such as those that exist for solving equations.

Algebraic word problems should not be read with the objective of immediately determining the answer because only in the simpler problems is this possible. Word problems should be initially read to identify what answer is asked for and to determine which quantity or quantities, if known, will give this answer. All of these quantities are called the unknowns in the problem. Recognizing all of the unknowns and writing algebraic expressions to describe them is often the most difficult part of solving word problems. Quite often, it is possible to identify and express the unknowns in several different ways and still solve the problem. Just as often, it is possible to identify and express the unknowns in several ways that appear different but are actually the same relationship.

In writing algebraic expressions for the various quantities given in word problems, it is helpful to look for certain words that indicate mathematical operations. The words "sum" and "total" signify addition; the word "difference" signifies subtraction; the words "product," "times," and "multiples of" signify multiplication; the words "quotient," "divided by," "per," and "ratio" signify division; and the words "same as" and "equal to" signify equality. When quantities are connected by these words and others like them, these quantities can be written as algebraic expressions.

Sometimes you may want to write equations initially using words. For example, Bob is 30 years older than Joe. Express Bob's age in terms of Joe's.

Bob's age = Joe's age plus 30 years

If we let Bob's age be represented by the symbol B and Joe's age by the symbol J, this becomes

$$B = J + 30 \text{ years}$$

Examples:

Equations:

1. The total electrical output of one nuclear facility is 200 megawatts more than that of another nuclear facility.

   Let L be the output of the larger facility and S the capacity of the smaller facility. The statement above written in equation form becomes $L = 200MW + S$.

2. The flow in one branch of a piping system is one-third that in the other branch.

   If B is the flow in the branch with more flow, and b is the flow in the smaller branch, this statement becomes the equation $b = \frac{1}{3} B$.

3. A man is three times as old as his son was four years ago.

   Let M = man's age and S = son's age. Then $M = 3\ (S\text{-}4)$.

4. A car travels in one hour 40 miles less than twice as far as it travels in the next hour.

   Let x_1 be the distance it travels the first hour and x_2 the distance it travels the second then, $x_1 = (2)\ (x_2)\ \text{-}40$.

Steps for Solving Algebraic Word Problems

Algebraic word problems can involve any number of unknowns, and they can require any number of equations to solve. However, regardless of the number of unknowns or equations involved, the basic approach to solving these problems is the same. First, condense the available information into algebraic equations, and, second, solve the equations. The most straightforward type of algebraic word problems are those that require only one equation to solve. These problems are solved using five basic steps.

Step 1. Let some letter, such as x, represent one of the unknowns.

Step 2. Express the other unknowns in terms of x using the information given in the problem.

Step 3. Write an equation that says in symbols exactly what the problem says in words.

Step 4. Solve the equation.

Step 5. Check the answer to see that it satisfies the conditions stated in the problem.

Example 1:

What are the capacities of two water storage tanks in a nuclear facility if one holds 9 gallons less than three times the other, and their total capacity is 63 gallons?

Solution:

Step 1. Let x = Capacity of the Smaller Tank

Step 2. Then, $3x - 9$ = Capacity of the Larger Tank

Step 3. Total Capacity = Capacity of the Smaller Tank + Capacity of the Larger Tank

$$63 = x + (3x - 9)$$

Step 4. Solving for x:

$$x + (3x - 9) = 63$$

$$4x - 9 = 63$$

$$4x = 63 + 9$$

$$4x = 72$$

$$x = 18$$

Solving for the other unknown:

$$3x - 9 = 3(18) - 9$$

$$3x - 9 = 54 - 9$$

$$3x - 9 = 45$$

Answer: Capacity of the Smaller Tank = 18 gallons
Capacity of the Larger Tank = 45 gallons

Step 5. The larger tank holds 9 gallons less than three times the smaller tank.

$$3(18) - 9 = 54 - 9 = 45$$

The total capacity of the two tanks is 63 gallons.

$$18 + 45 = 63$$

Thus, the answers check.

Example 2:

A utility has three nuclear facilities that supply a total of 600 megawatts (Mw) of electricity to a particular area. The largest facility has a total electrical output three times that of the smallest facility. The third facility has an output that is 50 Mw more than half that of the largest facility. What is the electrical output of each of the three facilities?

Solution:

Step 1. Let x = Electrical Output of the Smallest Facility.

Step 2. Then,

$3x$ = Electrical Output of the Largest Facility,

and,

$\frac{3x}{2} + 50$ = Electrical Output of the Third Facility.

Step 3. Total Electrical Output = Sum of the Electrical Outputs of the Three Facilities.

$$600 = x + 3x + \frac{3x}{2} + 50$$

Step 4. Solving for x:

$$x + 3x + \frac{3x}{2} + 50 = 600$$

$$\frac{2x}{2} + \frac{6x}{2} + \frac{3x}{2} = 600 - 50$$

$$\frac{11x}{2} = 550$$

$$11x = 1100$$

$$x = 100$$

Solving for the other unknowns:

$$3x = 3(100)$$

$$3x = 300$$

$$\frac{1}{2}(3x) + 50 = \frac{1}{2}(300) + 50$$

$$\frac{1}{2}(3x) + 50 = 150 + 50$$

$$\frac{1}{2}(3x) + 50 = 200$$

Answers: Electrical Output of the Smallest Facility = 100 Mw
Electrical Output of the Largest Facility = 300 Mw
Electrical Output of the Third Facility = 200 Mw

Step 5. The largest facility has a total electrical output three times that of the smallest facility.

$$3(100) = 300$$

The other facility has an output which is 50 Mw more than half that of the largest facility.

$$\frac{1}{2}(300) + 50 = 150 + 50 = 200$$

The total output of the three facilities is 600 Mw.

$$100 + 200 + 300 = 600$$

Thus, the answers check.

Example 3:

The winning team in a football game scored 7 points less than twice the score of the losing team. If the total score of both teams was 35 points, what was the final score?

Solution:

Step 1. Let x = Winning Team's Score

Step 2. Then, $\frac{1}{2}(x + 7)$ = Losing Team's Score

Step 3. Total Score = Winning Team's Score + Losing Team's Score

$$35 = x + \frac{1}{2}(x + 7)$$

Step 4. Solving for x:

$$x + \frac{1}{2}(x + 7) = 35$$

$$2x + x + 7 = 70$$

$$3x = 70 - 7$$

$$3x = 63$$

$$x = 21 \text{ points}$$

Solving for the other unknowns:

$$\frac{1}{2}(x + 7) = \frac{1}{2}(21 + 7)$$

$$\frac{1}{2}(x + 7) = \frac{1}{2}(28)$$

$$\frac{1}{2}(x + 7) = 14 \text{ points}$$

Answers: Winning Team's Score = 21 points
Losing Team's Score = 14 points

Step 5. The winning team's score is 7 points less than twice the score of the losing team.

$$2(14) - 7 = 28 - 7 = 21 \text{ points}$$

The total score of both teams is 35 points.

$$21 + 14 = 35 \text{ points}$$

Thus, the answers check.

Example 4:

A man is 21 years older than his son. Five years ago he was four times as old as his son. How old is each now?

Solution:

Step 1. Let x = Son's Age Now

Step 2. Then,

$x + 21$ = Father's Age Now
$x - 5$ = Son's Age Five Years Ago
$(x + 21) - 5$ = Father's Age Five Years Ago

Step 3. Five years ago the father was four times as old as his son.

$$(x + 21) - 5 = 4(x - 5)$$

Step 4.

$$\begin{aligned} (x + 21) - 5 &= 4(x - 5) \\ x + 16 &= 4x - 20 \\ x - 4x &= -20 - 16 \\ -3x &= -36 \\ x &= 12 \text{ years} \end{aligned}$$

Solving for the other unknowns:

$$\begin{aligned} x + 21 &= 12 + 21 \\ x + 21 &= 33 \text{ years} \end{aligned}$$

Answers: Son's Age Now = 12 years
Father's Age Now = 33 years

Step 5. The man is 21 years older than his son.

$$12 + 21 = 33 \text{ years}$$

Five years ago he was four times as old as his son.

$$33 - 5 = 28 = 4(12 - 5) = 4 \times 7$$

Thus, the answers check.

Word Problems Involving Money

The five basic steps for solving algebraic word problems can be used for solving word problems involving money. Writing algebraic expressions for these problems depends on the general relationship between the total value and the unit value of money. The total value of a collection of money or a collection of items with a certain monetary value equals the sum of the numbers of items each multiplied by their unit values. Thus, the total value of five pennies, three nickels, four dimes, and two quarters is found by solving the following equation:

$$x = 5(\$0.01) + 3(\$0.05) + 4(\$.10) + 2(\$0.25)$$

$$x = \$0.05 + \$0.15 + \$0.40 + \$0.50$$

$$x = \$1.10$$

The total value of 25 tickets worth \$1.50 each and 30 tickets worth \$0.75 each is 25(\$1.50) + 30(\$0.75) which equals \$37.50 + \$22.50 or \$60.00. Algebraic word problems involving money are solved using this general relationship following the same five basic steps for solving any algebraic word problems.

Example 1:

The promoter of a track meet engages a 6,000 seat armory. He wants to gross \$15,000. The price of children's tickets is to be one-half the price of adults' tickets. If one-third of the crowd is children, what should be the price of tickets, assuming capacity attendance?

Solution:

Step 1. Let x = Price of an Adult Ticket (in dollars)

Step 2. Then,

$\frac{x}{2}$ = Price of a Child's Ticket (in dollars)

$\frac{1}{3}(6{,}000) = 2{,}000$ = Number of Children's Tickets

$6{,}000 - 2{,}000 = 4{,}000$ = Number of Adults' Tickets

Step 3. Gross Income = (Number of Children's Tickets times their Unit Price) + (Number of Adults' Tickets times their Unit Price)

$$\$15{,}000 = 2{,}000\left(\frac{x}{2}\right) + 4{,}000(x)$$

Step 4. Solving for x:

$$15{,}000 = 2{,}000\left(\frac{x}{2}\right) + 4{,}000(x)$$

$$15{,}000 = 1{,}000x + 4{,}000x$$

$$15{,}000 = 5{,}000x$$

$$x = \$3.00$$

solving for the other unknown:

$\frac{x}{2}$ = Price of a Child's Ticket (in dollars)

$$\frac{x}{2} = \frac{\$3.00}{2}$$

$$\frac{x}{2} = \$1.50$$

Answers: Price of Adults' Tickets = \$3.00
Price of Children's Tickets = \$1.50

Step 5. The price of children's tickets is one-half the price of adults' tickets.

$$\frac{1}{2}(\$3.00) = \$1.50$$

The gross is \$15,000.

$4,000(\$3.00) + 2,000(\$1.50) = \$12,000 + \$3,000 = \$15,000$

Thus, the answers check.

Example 2:

A collection of coins consists of nickels, dimes, and quarters. The number of quarters is twice the number of nickels, and the number of dimes is five more than the number of nickels. If the total amount of money is \$5.05, how many of each type of coin are in the collection?

Solution:

Step 1. Let x = Number of Nickels

Step 2. Then,

$2x$ = Number of Quarters
$x + 5$ = Number of Dimes

Step 3. Total Value = (Number of Nickels)(Value of a Nickel) + (Number of Dimes)(Value of a Dime) + (Number of Quarters)(Value of a Quarter)

$\$5.05 = (x)(\$0.05) + (x + 5)(\$0.10) + (2x)(\$0.25)$

Step 4. Solving for x:

$\$5.05 = (x)(\$0.05) + (x + 5)(\$0.10) + (2x)(\$0.25)$

$\$5.05 = \$0.05x + \$0.10x + \$0.50 + \$0.50x$

$\$5.05 = \$0.65x + \$0.50$

$\$0.65x = \$5.05 - \$0.50$

$\$0.65x = \$4.55$

$$x = \frac{\$4.55}{\$0.65}$$

$$x = 7$$

Solving for the other unknowns:

$$2x = 2(7)$$

$$2x = 14$$

$$x + 5 = 7 + 5$$

$$x + 5 = 12$$

Answers: Number of Nickels = 7

Number of Dimes = 12

Number of Quarters = 14

Step 5. The number of quarters is twice the number of nickels.

$$2(7) = 14$$

The number of dimes is five more than the number of nickels.

$$7 + 5 = 12$$

The total value is \$5.05.

$$7(\$0.05) + 12(\$0.10) + 14(\$0.25) =$$

$$\$0.35 + \$1.20 + \$3.50 = \$5.05$$

Thus, the answers check.

Problems Involving Motion

Many algebraic word problems involve fundamental physical relationships. Among the most common are problems involving motion. For example, the definition of speed is distance traveled divided by the time it takes. $V_{ave} = \frac{\text{distance}}{\text{time}} = \frac{d}{t}$ or multiplying both sides by t, $d = V_{ave} \times t$. For example, if a car travels at 50 miles per hour for 2 hours, the distance traveled equals (50 mi/hr)(2 hr) or 100 miles. This relationship applies for constant velocity motion only. In practice, it is applied more generally by using an average speed or average rate of travel for the time involved. The distance traveled is often represented by s; the average speed or average rate of travel, also called the average velocity, by v_{av}; and the time of travel by t.

$$s = v_{av}t \qquad (2\text{-}13)$$

This same basic physical relationship can be written in two other forms, obtained by dividing both sides of the equation by v_{av} or by t.

$$t = \frac{s}{v_{av}} \qquad (2\text{-}14)$$

$$v_{av} = \frac{s}{t} \qquad (2\text{-}15)$$

Example 1:

How far can a car traveling at a rate of 52 miles per hour travel in 2½ hours?

Solution:

Using Equation 2-13:

$$s = v_{av}t$$

$$s = (52 \text{ miles/hour})(2\tfrac{1}{2} \text{ hours})$$

$$s = 130 \text{ miles}$$

Example 2:

How long does it take a plane traveling at 650 miles per hour to go 1430 miles?

Solution:

Using Equation 2-14:

$$t = \frac{s}{v_{av}}$$

$$t = \frac{1430 \text{ miles}}{650 \frac{\text{miles}}{\text{hour}}}$$

$$t = 2.2 \text{ hours}$$

Example 3:

What is the average speed of a train that completes a 450-mile trip in 5 hours?

Solution:

Using Equation 2-15:

$$v_{av} = \frac{s}{t}$$

$$v_{av} = \frac{450 \text{ miles}}{5 \text{ hours}}$$

$$v_{av} = 90 \text{ miles/hour}$$

Algebraic word problems involving motion are solved using the general relationship among distance, time, and average velocity following the same five basic steps for solving any algebraic word problem.

Example 1:

A plane flying at 525 miles per hour completes a trip in 2 hours less than another plane flying at 350 miles per hour. What is the distance traveled?

Solution:

Step 1. Let x = Distance Traveled (in miles)

Step 2. Then, using Equation 2-14,

$$\frac{x}{525} = \text{Time Taken by Faster Plane (in hours)}$$

$$\frac{x}{350} = \text{Time Taken by Slower Plane (in hours)}$$

Step 3. Time Taken by Faster Plane = Time Taken by Slower Plane - 2 hours

$$\frac{x}{525} \text{ hours} = \frac{x}{350} \text{ hours} - 2 \text{ hours}$$

$$\frac{x}{525} = \frac{x}{350} - \frac{700}{350}$$

$$\frac{x}{525} = \frac{x - 700}{350}$$

$$(350)(525)\left(\frac{x}{525}\right) = \left(\frac{x - 700}{350}\right)(350)(525)$$

$$350x = 525 \cdot (x - 700)$$

$$350x = 525x - 367{,}500$$

$$350x - 525x = -367{,}500$$

$$-175x = -367{,}5000$$

$$\frac{-175x}{-175} = \frac{-367{,}500}{-175}$$

$$x = 2100 \text{ miles}$$

Solving for the other unknowns:

$$\frac{x}{525} = \text{Time Taken by Faster Plane (in hours)}$$

$$\frac{x}{525} = \frac{2100}{525}$$

$$\frac{x}{525} = 4 \text{ hours}$$

$$\frac{x}{350} = \text{Time Taken by Slower Plane (in hours)}$$

$$\frac{x}{350} = \frac{2100}{350}$$

$$\frac{x}{350} = 6 \text{ hours}$$

Answers: Distance Traveled = 2100 miles
Time Taken by Faster Plane = 4 hours
Time Taken by Slower Plane = 6 hours

Step 5. The faster plane takes 2 hours less to complete the trip than the slower plane.

6 hours - 2 hours = 4 hours

Thus, the answer checks.

Example 2:

It takes a man 4 hours to reach a destination 1325 miles from his home. He drives to the airport at an average speed of 50 miles per hour, and the average speed of his plane trip is 500 miles per hour. How far does he travel by each mode of transportation?

Solution:

Step 1. Let x = Distance Traveled by Car (in miles)

Step 2. Then,

1325 - x = Distance Traveled by Plane (in miles)
and, using Equation 2-14,

$$\frac{x}{50} = \text{Time Traveled by Car (in hours)}$$

$$\frac{1325 - x}{500} = \text{Time Traveled by Plane (in hours)}$$

Step 3. Total Time = (Time Traveled by Car) + (Time Traveled by Plane)

$$4 \text{ hours} = \frac{x}{50} \text{ hours} + \frac{1325 - x}{500} \text{ hours}$$

Step 4. Solving for x:

$$4 = \frac{x}{50} + \frac{1325 - x}{500}$$

$$4 = \frac{10x + 1325 - x}{500}$$

$$(500)4 = \frac{9x + 1325}{500}(500)$$

$$2000 = 9x + 1325$$

$$2000 - 1325 = 9x$$

$$685 = 9x$$

$$\frac{9x}{9} = \frac{675}{9}$$

$$x = 75 \text{ miles}$$

Solving for the other unknowns:

$$\frac{x}{50} = \text{Time Traveled by Car (in hours)}$$

$$\frac{x}{50} = \frac{75}{50}$$

$$\frac{x}{50} = 1\frac{1}{2} \text{ hours}$$

$$\frac{1325 - x}{500} = \text{Time Traveled by Plane (in hours)}$$

$$\frac{1325 - x}{500} = \frac{1324 - 75}{500}$$

$$\frac{1325 - x}{500} = \frac{1250}{500} = 2\frac{1}{2} \text{ hours}$$

$$1325 - x = \text{Distance Traveled by Plane (in miles)}$$

$$1325 - x = 1325 - 75$$

$$1325 - x = 1250 \text{ miles}$$

Answers: Distance Traveled by Car = 75 miles
Distance Traveled by Plane = 1250 miles

Step 5. The total distance traveled is 1325 miles.

75 miles + 1250 miles = 1325 miles

The average speed by car is 50 miles per hour.

$$\frac{75 \text{ miles}}{1\frac{1}{2} \text{ hours}} = 50 \text{ miles per hour}$$

The average speed by plane is 500 miles per hour.

$$\frac{1250 \text{ miles}}{2\frac{1}{2} \text{ hours}} = 500 \text{ miles per hour}$$

The total time traveling is 4 hours.

$1\frac{1}{2}$ hours + $2\frac{1}{2}$ hours = 4 hours

Thus, the answers check.

Solving Word Problems Involving Quadratic Equations

Many algebraic word problems involve quadratic equations. Any time the algebraic expressions describing the relationships in the problem involve a quantity multiplied by itself, a quadratic equation must be used to solve the problem. The steps for solving word problems involving quadratic equations are the same as for solving word problems involving linear equations.

Example:

A radiation control point is set up near a solid waste disposal facility. The pad on which the facility is set up measures 20 feet by 30 feet. If the health physicist sets up a controlled walkway around the pad that reduces the area by 264 square feet, how wide is the walkway?

Solution:

Step 1. Let x = Width of the Walkway

Step 2. Then,
$30 - 2x$ = Length of Reduced Pad
$20 - 2x$ = Width of Reduced Pad

Step 3. Area of Reduced Pad = (Length of Reduced Pad)(Width of Reduced Pad)

$$600 - 264 = (30 - 2x)(20 - 2x)$$

$$336 = 600 - 100x + 4x^2$$

Step 4. Solve this quadratic equation.

$$4x^2 - 100x + 264 = 0$$

Using the Quadratic Formula, substitute the coefficients for *a*, *b*, and *c* and solve for *x*.

$$x = \frac{-b \pm \sqrt{b^2 - 4ac}}{2a}$$

$$x = \frac{-(-100) \pm \sqrt{(-100)^2 - 4(4)(264)}}{2(4)}$$

$$x = \frac{100 \pm \sqrt{10{,}000 - 4{,}224}}{8}$$

$$x = \frac{100 \pm \sqrt{5{,}776}}{8}$$

$$x = \frac{100 \pm 76}{8}$$

$$x = \frac{100 + 76}{8}, \frac{100 - 76}{8}$$

$$x = \frac{176}{8}, \frac{24}{8}$$

$$x = 22,\ 3$$

The two roots are $x = 22$ feet and $x = 3$ feet. Since $x = 22$ feet is not physically meaningful, the answer is $x = 3$ feet.

Step 5. Check the answer.

The area of the reduced area pad is 264 square feet less than the area of the original pad.

$$600 - 264 = (20 - 2x)(30 - 2x)$$

$$336 = [20 - 2(3)][30 - 2(3)]$$

$$336 = (20 - 6)(30 - 6)$$

$$336 = (14)(24)$$

$$336 = 336$$

Thus, the answer checks.

Summary

The important information from this chapter is summarized below.

Algebraic Word Problems Summary

Algebraic word problems can easily be solved by following these five basic steps:

Step 1. Let some letter, such as x, represent one of the unknowns.

Step 2. Express the other unknowns in terms of x using the information given in the problem.

Step 3. Write an equation that represents in symbols exactly what the problem states in words.

Step 4. Solve the equation.

Step 5. Check the answer to see that it satisfies the conditions stated in the problem.

LOGARITHMS

This chapter covers changing the base of a logarithm and solving problems with logarithms.

EO 1.6 STATE the definition of a logarithm.

EO 1.7 CALCULATE the logarithm of a number.

Calculator Usage, Special Keys

This chapter will require the use of certain keys on a calculator to perform the necessary calculations. An understanding of the functions of each key will make logarithms (logs) an easy task.

Common Logarithm key

[log] This key when pressed will compute the common log (base 10) of the number x in the display, where x is greater than zero.

Natural Logarithm key

[ln x] This key when pressed will compute the natural logarithm (base e) of the number x in the display, where x is greater than zero.

[INV] This key when pressed before the log and ln keys will compute the anti-log of the number x in the display. When used with the log key it will raise 10 to the displayed power ($10^{7.12}$) and when used with the ln key will raise (e) to the displayed power (e^{-381}).

Introduction

Logarithms are exponents, as will be explained in the following sections. Before the advent of calculators, logarithms had great use in multiplying and dividing numbers with many digits since adding exponents was less work than multiplying numbers. Now they are important in nuclear work because many laws governing physical behavior are in exponential form. Examples are radioactive decay, gamma absorption, and reactor power changes on a stable period.

Definition

Any number (X) can be expressed by any other number b (except zero) raised to a power x; that is, there is always a value of x such that $X = b^x$. For example, if $X = 8$ and $b = 2$, $x = 3$. For $X = 8$ and $b = 4$, $8 = 4^x$ is satisfied if $x = 3/2$.

$$4^{\frac{3}{2}} = (4^3)^{\frac{1}{2}} = (64)^{\frac{1}{2}} = 8$$

or

$$4^{\frac{3}{2}} = (4^{\frac{1}{2}})^3 = 2^3 = 8$$

In the equation $X = b^x$, the exponent x is the logarithm of X to the base b. Stated in equation form, $x = \log_b X$, which reads x is the logarithm to the base b of X. In general terms, the logarithm of a number to a base b is the power to which base b must be raised to yield the number. The rules for logs are a direct consequence of the rules for exponents, since that is what logs are. In multiplication, for example, consider the product of two numbers X and Y. Expressing each as b raised to a power and using the rules for exponents:

$$XY = (b^x)\,(b^y) = b^{x+y}$$

Now, equating the $\log_b$ of the first and last terms, $\log_b XY = \log_b b^{x+y}$.

Since the exponent of the base b $(x+y)$ is the logarithm to the base b, $\text{Log}_b\, b^{x+y} = x+y$.

$$\log_b XY = x+y$$

Similarily, since $X = b^x$ and $Y = b^y$, $\log_b X = x$ and $\log_b Y = y$. Substituting these into the previous equation,

$$\log_b XY = \log_b X + \log_b Y$$

Before the advent of hand-held calculators it was common to use logs for multiplication (and division) of numbers having many significant figures. First, logs for the numbers to be multiplied were obtained from tables. Then, the numbers were added, and this sum (logarithm of the product) was used to locate in the tables the number which had this log. This is the product of the two numbers. A slide rule is designed to add logarithms as numbers are multiplied.

Logarithms can easily be computed with the calculator using the keys identified earlier.

Examples:

$\log_2 8 = 3$ since $8 = 2^3$

$\log_{10} 0.01 = -2$ since $0.01 = 10^{-2}$

$\log_5 5 = 1$ since $5 = 5^1$

$\log_b 1 = 0$ since $1 = b$

From the above illustration, it is evident that a logarithm is an exponent. 3^4 is called the exponential form of the number 81. In logarithmic form, 3^4 would be expressed as $\log_3 81 = 4$, or the logarithm of 81 to the base 3 is 4. Note the symbol for taking the logarithm of the number 81 to a particular base 3, is $\log_3 81$, where the base is indicated by a small number written to the right and slightly below the symbol log.

Log Rules

Since logs are exponents, the rules governing logs are very similar to the laws of exponents.

The most common log rules are the following:

1. $\log_b (ABC) = \log_b A + \log_b B + \log_b C$
2. $\log_b (A/B) = \log_b A - \log_b B$
3. $\log_b (A^n) = n\log_b A$
4. $\log_b b = 1$
5. $\log_b \sqrt[n]{A} = \log_b A^{1/n} = (1/n)\log_b A$
6. $\log_b 1 = 0$
7. $\log_b (1/A) = \log_b 1 - \log_b A = -\log_b A$

Example 1: $y = \frac{1}{2} gt^2$ where $g = 32$

Solution:

$y = 16\ t^2$

Find y for $t = 10$ using logs.

$\log_{10} y = \log_{10} (16\ t^2)$

$\log_{10} y = \log_{10} 16 + \log_{10} t^2$

$\log_{10} y = \log_{10} 16 + (2 \log_{10} t)$

$\log_{10} y = 1.204 + 2 \log_{10} 10$

$\log_{10} y = 1.204 + 2 \times 1$

$\log_{10} y = 3.204$

but this means $10^{3.204} = y$

$y = 1600$

Example 2: Calculate $\log_{10} 2 - \log_{10} 3$.

Solution:

Rule 2. $\log_{10} (A/B)$: $\log_{10} A - \log_{10} B$

$\log_{10} 2 - \log_{10} 3$

$= \log_{10} (2/3)$

$= \log_{10} (.667)$

$= -0.176$

Example 3: Calculate $3\log_{10} 2$.

Solution:

Rule 3. $\log_b (A^n) = n\log_b A$

$3\log_{10} 2$

$= \log_{10} (2^3)$

$= \log_{10} 8$

$= 0.903$

Example 4: Calculate $4\log_{10} 10$.

Solution:

Rule 4. $\log_b b = 1$

$4\log_{10} 10$

$= 4(1)$

$= 4$

Example 5: Calculate $(1/3)\log_{10} 2$.

Solution:

Rule 5. $\log_b \sqrt[n]{A} = \log_b A^{1/n} = (1/n)\log_b A$

$(1/3)\log_{10} 2$

$= \log_{10} \sqrt[3]{2}$

$= \log_{10} 1.259$

$= 0.1003$

Example 6: Calculate $\log_{10} 1$.

Solution:

Rule 6. $\log_b 1 = 0$

$\log_{10} 1 = 0$

Example 7: Calculate $-\log_{10} 2$.

Solution:

Rule 7: $\log_b (1/A) = -\log_b A$

$-\log_{10} 2$

$= \log_{10} (1/2)$

$= -\log_{10} 0.5$

$= -0.3010$

Common and Natural Logarithms

In scientific and engineering practice, the natural system of logarithms uses the number 2.718281828459042. Since this number is frequently encountered, the letter *e* is used. Many natural occurrences can be expressed by exponential equations with *e* as the base. For example, the decay of radioactive isotopes can be expressed as a natural logarithm equation. These logarithmic expressions are called natural logs because *e* is the basis for many laws of nature.

The expression *ln* is used to represent a logarithm when *e* is the base. Therefore, the exponential equation is written as

$$e^x = N.$$

and the logarithm expression is written as

$$\log_e N = x \quad \text{or} \quad \ln N = x.$$

As with base 10 logs (common logs), natural logs can be determined easily with the aid of a calculator.

Base 10 logs are often referred to as common logs. Since base 10 is the most widely used number base, the "10" from the designation $\log_{10}$ is often dropped. Therefore, any time "log" is used without a base specified, one should assume that base 10 is being used.

Anti-Logarithms

An anti-logarithm is the opposite of a logarithm. Thus, finding the anti-logarithm of a number is the same as finding the value for which the given number is the logarithm. If $\log_{10} X = 2$, then 2.0 is the power (exponent) to which one must raise the base 10 to obtain X, that is, $X = 10^{2.0} = 100$. The determination of an anti-log is the reverse process of finding a logarithm.

Example:

Multiply 38.79 and 6896 using logarithms.

Log 38.79 = 1.58872 Log 6896 = 3.83860

Add the logarithms to get 5.42732

Find the anti-log.

Anti-log 5.42732 = 2.675×10^5 = 267,500

Thus, 38.79 x 6896 = 2.675×10^5 = 267,500

Natural and Common Log Operations

The utilization of the log/ln can be seen by trying to solve the following equation algebraically. This equation cannot be solved by algebraic methods. The mechanism for solving this equation is as follows:

<u>Using Common Logs</u>

$$2^X = 7$$

$$\log 2^X = \log 7$$

$$X \log 2 = \log 7$$

$$X = \frac{\log 7}{\log 2} = \frac{0.8451}{0.3010} = 2.808$$

<u>Using Natural Logs</u>

$$2^X = 7$$

$$\ln 2^X = \ln 7$$

$$X \ln 2 = \ln 7$$

$$X = \frac{\ln 7}{\ln 2} = \frac{1.946}{0.693} = 2.808$$

How would you calculate x in the following equation?

$$\log x = 5$$

The easy way to solve this equation is to take the anti-log. As division is the reverse of multiplication, so anti-log is the reverse of log. To take the anti-log $\log_{10} x = 5$:

$$\text{anti-log} (\log X) = \text{anti-log } 5$$

$$x = \text{anti-log } 5$$

$$x = 100{,}000$$

This is accomplished on a calculator by pressing the 5, INV, then the LOG key. This causes the inverse of the log process.

Summary

The important information in this chapter is summarized below.

Logarithms Summary

- A number L is said to be the logarithm of a positive real number N to the base b (where b is real, positive, and not equal to 1), if L is the exponent to which b must be raised to obtain N, or the function can be expressed as

  $$L = \text{Log}_b N$$

  for which the inverse is

  $$N = b^L$$

  Simply stated, the logarithm is the inverse of the exponential function.

- Product = base^{exponent}
- $\text{Log}_{\text{base}}\text{product} = \text{exponent}$
- $\text{Log}_b (ABC) = \log_b A + \log_b B + \log_b C$
- $\text{Log}_b (A/B) = \log_b A - \log_b B$
- $\text{Log}_b (A^n) = n\log_b A$
- $\text{Log}_b \sqrt[n]{A} = \log_b A^{1/n} = (1/n)\log_b A$
- $\text{Log}_b 1 = 0$
- $\text{Log}_b (1/A) = \log_b 1 - \log_b A = -\log_b A$
- Common logs are base 10
- Natural logs are base e
- Anti-log is the opposite of a log

GRAPHING

This chapter covers graphing functions and linear equations using various types of graphing systems.

EO 1.8 **STATE the definition of the following terms:**
- **a. Ordinate**
- **b. Abscissa**

EO 1.9 **Given a table of data, PLOT the data points on a cartesian coordinate graph.**

EO 1.10 **Given a table of data, PLOT the data points on a logarithmic coordinate graph.**

EO 1.11 **Given a table of data, PLOT the data points on the appropriate graphing system to obtain the specified curve.**

EO 1.12 **Obtain data from a given graph.**

EO 1.13 **Given the data, SOLVE for the unknown using a nomograph.**

In work with physical systems, the relationship of one physical quantity to another is often of interest. For example, the power level of a nuclear reactor can be measured at any given time. However, this power level changes with time and is often monitored. One method of relating one physical quantity to another is to tabulate measurements. Thus, the power level of a nuclear reactor at specific times can be recorded in a log book. Although this method does provide information on the relationship between power level and time, it is a difficult method to use effectively. In particular, trends or changes are hard to visualize. Graphs often overcome these disadvantages. For this reason, graphs are widely used.

A graph is a pictorial representation of the relationship between two or more physical quantities. Graphs are used frequently both to present fundamental data on the behavior of physical systems and to monitor the operation of such systems. The basic principle of any graph is that distances are used to represent the magnitudes of numbers. The number line is the simplest type of graph. All numbers are represented as distances along the line. Positive numbers are located to the right of zero, and negative numbers are located to the left of zero.

The coordinate system of a graph is the framework upon which the graph is drawn. A coordinate system consists of numbered scales that give the base and the direction for measuring points on the graph. Any point on a graph can be specified by giving its coordinates. Coordinates describe the location of the point with respect to the scales of the coordinate system. There are several different coordinate systems commonly encountered.

The Cartesian Coordinate System

The Cartesian Coordinate System, also known as the rectangular coordinate system, consists of two number scales, called the x-axis (at y = 0) and the y-axis (at x = 0), that are perpendicular to each other. Each scale is a number line drawn to intersect the other at zero. The zero point is called the origin. The divisions along the scales may be any size, but each division must be equal. Figure 1 shows a rectangular coordinate system. The axes divide the coordinate system into four regions called quadrants. Quadrant I is the region above the x-axis and to the right of the y-axis. Quadrant II is the region above the x-axis and to the left of the y-axis. Quadrant III is the region below the x-axis and to the left of the y-axis. Quadrant IV is the region below the x-axis and to the right of the y-axis.

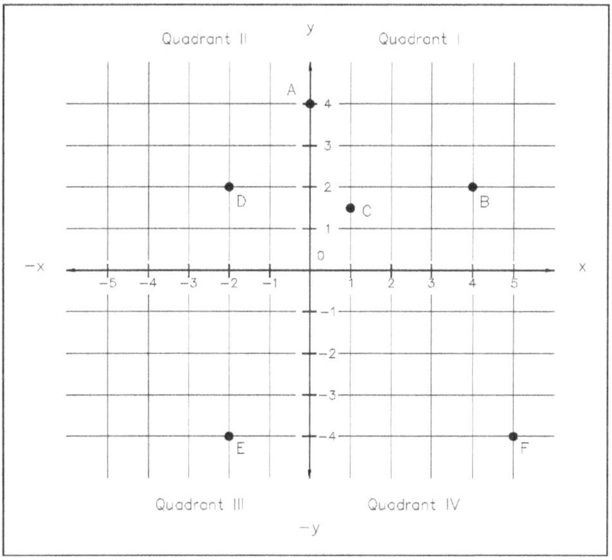

Figure 1 The Cartesian System

The use of a graph starts with the plotting of data points using the coordinate system. These data points are known as the **abscissa** and the **ordinate**. The abscissa, also known as the y-coordinate, is the distance along the y-axis. The ordinate, also known as the x-coordinate, is the distance along the x-axis. A point on a Cartesian coordinate graph is specified by giving its x-coordinate and its y-coordinate. Positive values of the x-coordinate are measured to the right, negative values to the left. Positive values of the y-coordinate are measured up, negative values down. For example, the x- and y-coordinates are both zero at the origin. The origin is denoted as (0,0), where the first zero refers to the value of the x-coordinate. Point A in Figure 1 is denoted as (0,4), since the value of the x-coordinate is zero, and the value of the y-coordinate is 4. In Quadrant I, every point has a positive x-coordinate and a positive y-coordinate. Point B in Figure 1 is located in Quadrant I and is denoted by (4,2). Fractional values of coordinates can also be shown. Point C in Figure 1 is denoted by (1,1.5). In Quadrant II, every point has a negative x-coordinate and a positive y-coordinate. Point D is denoted by (-2,2). In Quadrant III, every point has a negative x-coordinate and a negative y-coordinate. Point E is located in Quadrant III and is denoted by (-2,-4). In Quadrant IV, every point has a positive x-coordinate, but a negative y-coordinate. Point F is located in Quadrant IV and is denoted by (5,-4).

Cartesian Coordinate Graphs

The most common type of graph using the Cartesian Coordinate System is one in which all values of both the x-coordinate and the y-coordinate are positive. This corresponds to Quadrant I of a Cartesian coordinate graph. The relationship between two physical quantities is often shown on this type of rectangular plot. The x-axis and the y-axis must first be labeled to correspond to one of the physical quantities. The units of measurement along each axis must also be established. For example, to show the relationship between reactor power level and time, the x-axis can be used for time in minutes and the y-axis for the reactor power level as a percentage of full power level. Data points are plotted using the associated values of the two physical quantities.

Example: The temperature of water flowing in a high pressure line was measured at regular intervals. Plot the following recorded data on a Cartesian coordinate graph.

Time (min)	Temperature (°F)
0	400°
15	420°
30	440°
45	460°
60	480°
75	497°
90	497°
105	497°
120	497°

The first step is to label the x-axis and the y-axis. Let the x-axis be time in minutes and the y-axis be temperature in °F.

The next step is to establish the units of measurement along each axis. The x-axis must range from 0 to 120, the y-axis from 400 to 500.

The points are then plotted one by one. Figure 2 shows the resulting Cartesian coordinate graph.

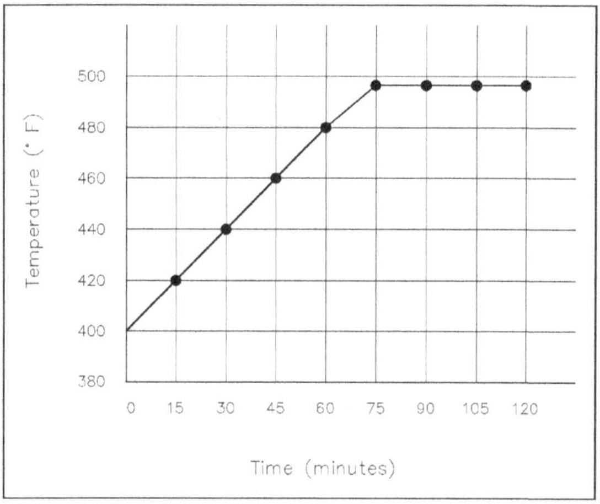

Figure 2 Cartesian Coordinate Graph of Temperature vs. Time

Example: The density of water was measured over a range of temperatures. Plot the following recorded data on a Cartesian coordinate graph.

Temperature (°C)	Density (g/ml)
40°	0.992
50°	0.988
60°	0.983
70°	0.978
80°	0.972
90°	0.965
100°	0.958

The first step is to label the x-axis and the y-axis. Let the x-axis be temperature in °C and the y-axis be density in g/ml.

The next step is to establish the units of measurement along each axis. The x-axis must range from approximately 40 to 100, the y-axis from 0.95 to 1.00.

The points are then plotted one by one. Figure 3 shows the resulting Cartesian coordinate graph.

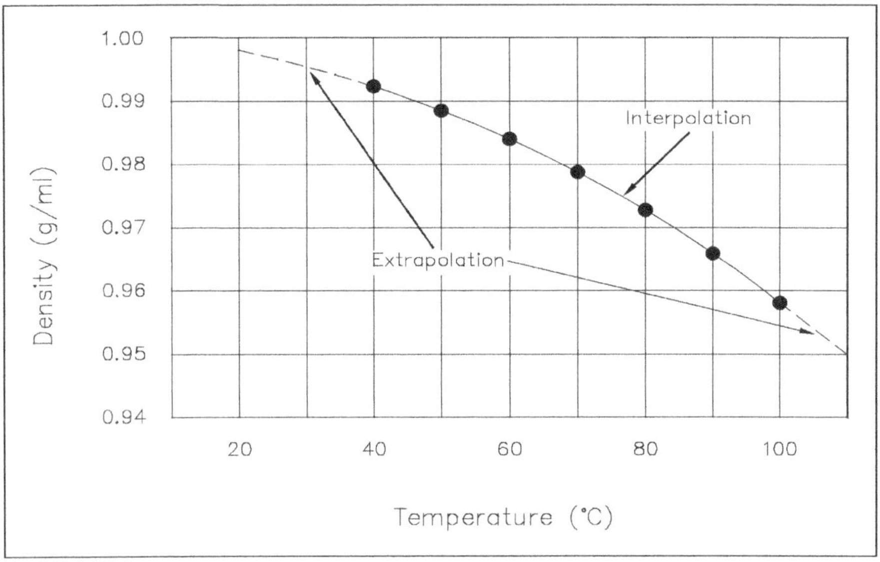

Figure 3 Cartesian Coordinate Graph of Density of Water vs. Temperature

Graphs are convenient because, at a single glance, the major features of the relationship between the two physical quantities plotted can be seen. In addition, if some previous knowledge of the physical system under consideration is available, the numerical value pairs of points can be connected by a straight line or a smooth curve. From these plots, the values at points not specifically measured or calculated can be obtained. In Figures 2 and 3, the data points have been connected by a straight line and a smooth curve, respectively. From these plots, the values at points not specifically plotted can be determined. For example, using Figure 3, the density of water at 65°C can be determined to be 0.98 g/ml. Because 65°C is within the scope of the available data, it is called an interpolated value. Also using Figure 3, the density of water at 101°C can be estimated to be 0.956 g/ml. Because 101°C is outside the scope of the available data, it is called an extrapolated value. Although the value of 0.956 g/ml appears reasonable, an important physical fact is absent and not predictable from the data given. Water boils at 100°C at atmospheric pressure. At temperatures above 100°C it is not a liquid, but a gas. Therefore, the value of 0.956 g/ml is of no significance except when the pressure is above atmospheric.

This illustrates the relative ease of interpolating and extrapolating using graphs. It also points out the precautions that must be taken, namely, interpolation and extrapolation should be done only if there is some prior knowledge of the system. This is particularly true for extrapolation where the available data is being extended into a region where unknown physical changes may take place.

Logarithmic Graphs

Frequently, the function to be plotted on a graph makes it convenient to use scales different from those used for the Cartesian coordinate graphs. Logarithmic graphs in which one or both of the scales are divided logarithmically are common. A semi-log plot is used when the function is an exponential, such as radioactive decay. A semi-log plot is obtained by using an ordinary linear scale for one axis and a logarithmic scale for the other axis. A log-log plot is used when the function is a power. A log-log plot is obtained by using logarithmic scales for both axes. Table 1 gives data on the amount of radioactive strontium 90 present as a function of time in years. Every twenty-five years one-half of the material decays. Figure 4 is a Cartesian coordinate graph of the data given in Table 1. It can be seen from Figure 4 that it is difficult to determine from this plot the amount of strontium 90 present after long periods of time such as 125 years, 150 years, or 175 years.

TABLE 1
Data on the Radioactive Decay of Strontium 90

Time (years)	Amount of Strontium 90 (grams)
0	100
25	50
50	25
75	12.5
100	6.25
125	3.125
150	1.5625
175	0.78125

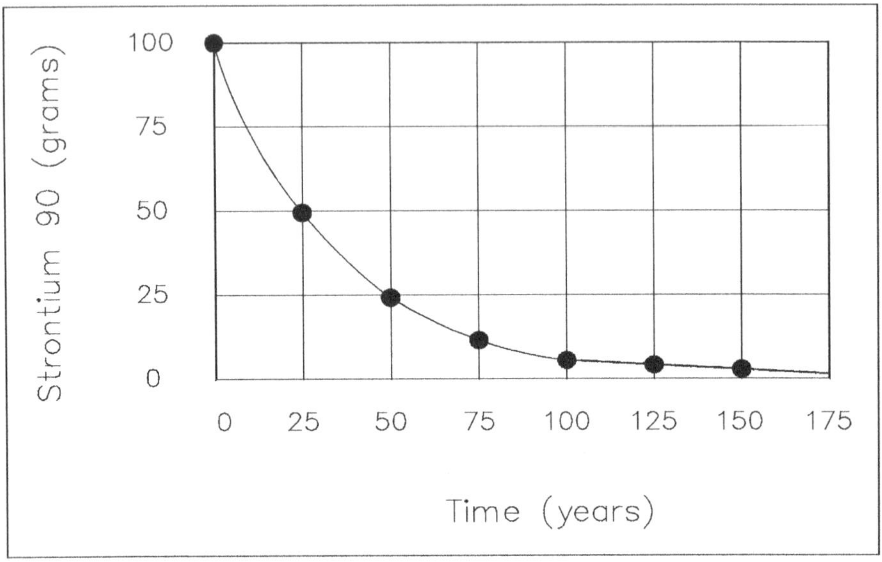

Figure 4 Cartesian Coordinate Plot of Radioactive Decay of Strontium 90

If the same data, the decay of strontium 90, is plotted on semi-log, the resulting plot (Figure 5) will be a straight line. This is because the decay of radioactive material is an exponential function. The resulting straight line of the semi-log plot allows a more accurate extrapolation or interpolation of the data than the curve obtained from the cartesian plot.

For graphs in which both of the quantities (x,y) vary as a power function, a log-log plot is convenient. A log-log plot is obtained by using logarithmic scales for both axes. Table 2 gives data on the frequency of electromagnetic radiation as a function of the wavelength of the radiation. Figure 6 is a log-log plot of the data given in Table 2.

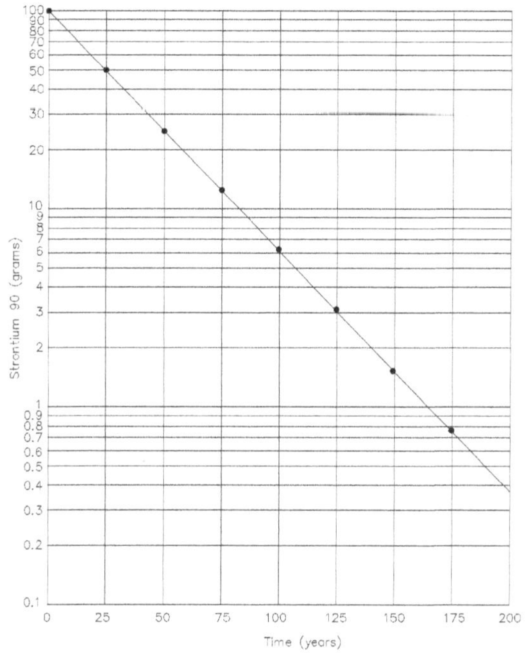

Figure 5 Semi-log Plot of Radioactive Decay of Strontium 90

TABLE 2
Data on Frequency vs. Wavelength of Electromagnetic Radiation

Wavelength (cm)	Frequency (s^{-1})
1.0×10^{-8}	3×10^{18}
0.5×10^{-7}	6×10^{17}
1.0×10^{-7}	3×10^{17}
0.5×10^{-6}	6×10^{16}
1.0×10^{-6}	3×10^{16}

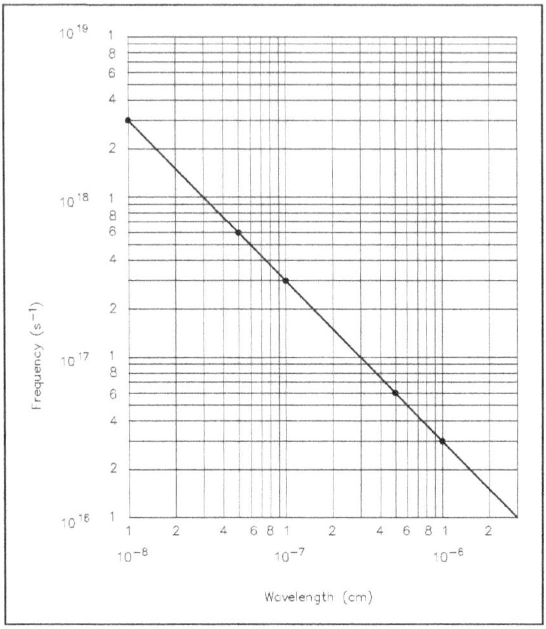

Figure 6 Log-Log Plot of Frequency vs. Wavelength of Electromagnetic Radiation

In summary, the type of coordinate system used to plot data, cartesian, semi-log, or log-log, should be based on the type of function to be graphed and the desired shape (curve or line) of the curve wanted.

- Cartesian system - Linear ($y = mx + b$) type functions when plotted will provide straight lines; exponential functions ($y = e^x$) will plot as curves.
- Semi-log system - Should not plot linear type functions on semi-log. Exponential functions, such as radioactive decay and reactor power equations when plotted will graph as straight lines.
- Log-log - Rarely used; used to plot power equations.

Graphing Equations

Algebraic equations involving two unknowns can readily be shown on a graph. Figure 7 shows a plot of the equation $x + y = 5$. The equation is solved for corresponding sets of values of x and y that satisfy the equation. Each of these points is plotted and the points connected. The graph of $x + y = 5$ is a straight line.

x	y
−3	8
−2	7
−1	6
0	5
1	4
2	3
3	2
4	1
5	0
6	−1
7	−2
8	−3

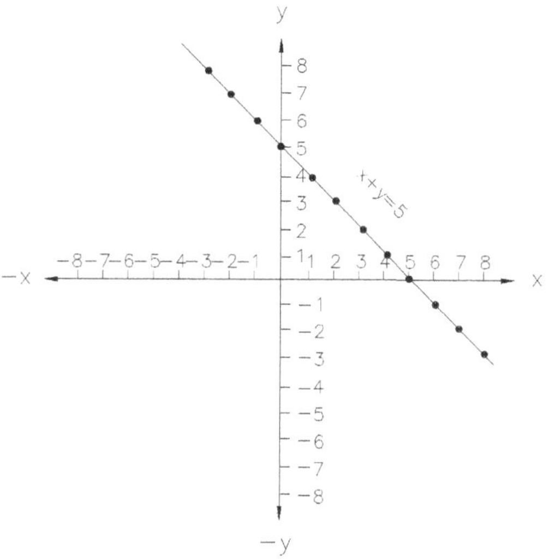

Figure 7 Plot of x + y = 5

The x-intercept of a line on a graph is defined as the value of the x-coordinate when the y-coordinate is zero. It is the value of x where the graph intercepts the x-axis. The y-intercept of a graph is defined as the value of the y-coordinate when the x-coordinate is zero. It is the value of y where the graph intercepts the y-axis. Thus, the x-intercept of the graph of $x + y = 5$ is +5. For a linear equation in the general form $ax + by = c$, the x-intercept and y-intercept can also be given in general form.

Any algebraic equation involving two unknowns of any function relating two physical quantities can be plotted on a Cartesian coordinate graph. Linear equations or linear functions plot as straight lines on Cartesian coordinate graphs. For example, $x + y = 5$ and $f(x) = 3x + 9$ plot as straight lines. Higher order equations or functions, such as quadratic equations or functions and exponential equations, can be plotted on Cartesian coordinate graphs. Figure 8 shows the shape of the graph of a typical quadratic equation or function. This shape is called a parabola. Figure 9 shows the shape of the graph of a typical exponential equation or function.

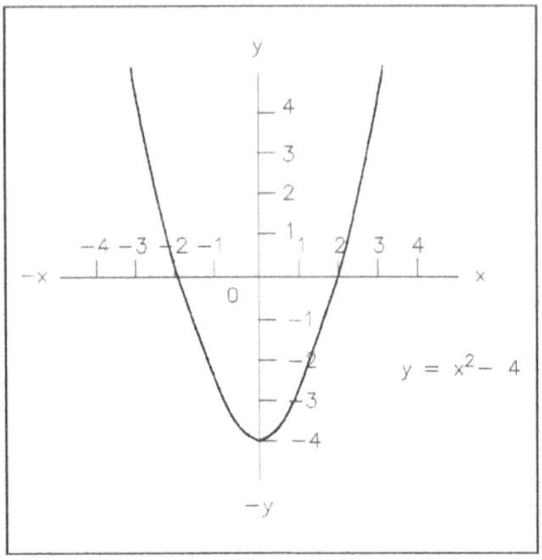

Figure 8 Cartesian Coordinate Graph of Quadratic Equation or Function

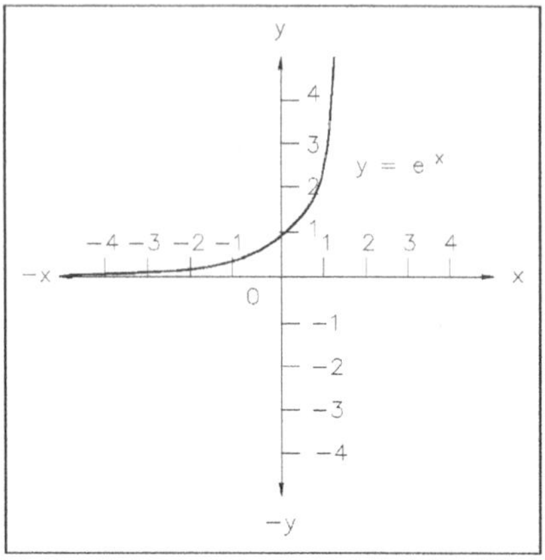

Figure 9 Cartesian Coordinate Graph of Exponential Equation or Function

Nomographs

A nomograph is a device used to relate the physical quantities in such a way that the value of an unknown quantity can be determined given the values of the other related quantities. Nomographs normally involve the relationship among three physical quantities. The scales are located in such a way that, when a straight line is drawn between the values of the known quantities on their respective scales, the line crosses the value of the unknown quantity on its scale. Figure 10 is a typical nomograph that relates the distance traveled, the average speed, and the time traveled. It should be noted that, as with any graphical representation, the values determined are only approximations.

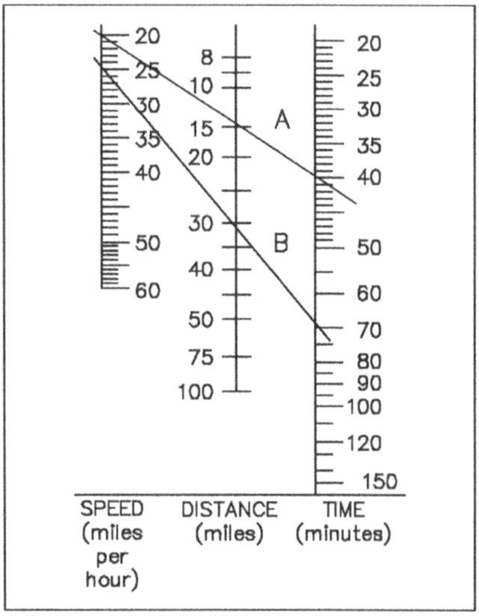

Figure 10 Typical Nomograph

Example:

Using Figure 10, find the distance traveled if the average speed is 20 mph and the time traveled is 40 minutes.

The line labeled A in Figure 10 connects 20 mph and 40 minutes. It passes through 14.5 miles.

Thus, the distance traveled is 14.5 miles.

Example:

Using Figure 10, find the time required to travel 31 miles at an average speed of 25 mph.

The line labeled B in Figure 10 connects 31 miles and 25 mph. It passes through 70 minutes.

Thus, the time required is 70 minutes.

Summary

The important information in this chapter is summarized below.

Graphing Summary

- Ordinate - x-coordinate
- Abscissa - y-coordinate
- Cartesian Coordinate System
  - Rectangular Coordinate System
  - Divided into four quadrants by x- and y-axis
- Logarithmic Coordinate System
  - One or both of the scales are divided logarithmically
  - Semi-log graphs contain linear x-axis and logarithmic y-axis
  - Log-log graphs contain logarithmic x- and y-axis
- Linear functions are usually plotted on Cartesian coordinate graph.
- Exponential functions ($y = e^x$) are usually plotted on semi-log graphs to provide a straight line instead of the resulting curve placed on a Cartesian coordinate graph.
- Power functions ($Y = ax^2$, $y = ax^3$, etc.) are usually plotted on log-log graphs.

SLOPES

This chapter covers determining and calculating the slope of a line.

EO 1.14 STATE the definition of the following terms:
a. Slope
b. Intercept

EO 1.15 Given the equation, CALCULATE the slope of a line.

EO 1.16 Given the graph, DETERMINE the slope of a line.

Many physical relationships in science and engineering may be expressed by plotting a straight line. The slope(m), or steepness, of a straight line tells us the amount one parameter changes for a certain amount of change in another parameter.

Slope

For a straight line, slope is equal to rise over run, or

$$\text{slope} = \frac{\text{rise}}{\text{run}} = \frac{\text{change in } y}{\text{change in } x} = \frac{\Delta y}{\Delta x} = \frac{y_2 - y_1}{x_2 - x_1}$$

Consider the curve shown in Figure 11. Points $P1$ and $P2$ are any two different points on the line, and a right triangle is drawn whose legs are parallel to the coordinate axes. The length of the leg parallel to the x-axis is the difference between the x-coordinates of the two points and is called "Δx," read "delta x," or "the change in x." The leg parallel to the y-axis has length Δy, which is the difference between the y-coordinates. For example, consider the line containing points (1,3) and (3,7) in the second part of the figure. The difference between the x-coordinates is $\Delta x = 3\text{-}1 = 2$. The difference between the y-coordinates is $\Delta y = 7\text{-}3 = 4$. The ratio of the differences, $\Delta y/\Delta x$, is the slope, which in the preceding example is 4/2 or 2. It is important to notice that if other points had been chosen on the same line, the ratio $\Delta y/\Delta x$ would be the same, since the triangles are clearly similar. If the points (2,5) and (4,9) had been chosen, then $\Delta y/\Delta x = (9\text{-}5)/(4\text{-}2) = 2$, which is the same number as before. Therefore, the ratio $\Delta y/\Delta x$ depends on the inclination of the line, m = rise [vertical (y-axis) change] ÷ run [horizontal (x-axis) change].

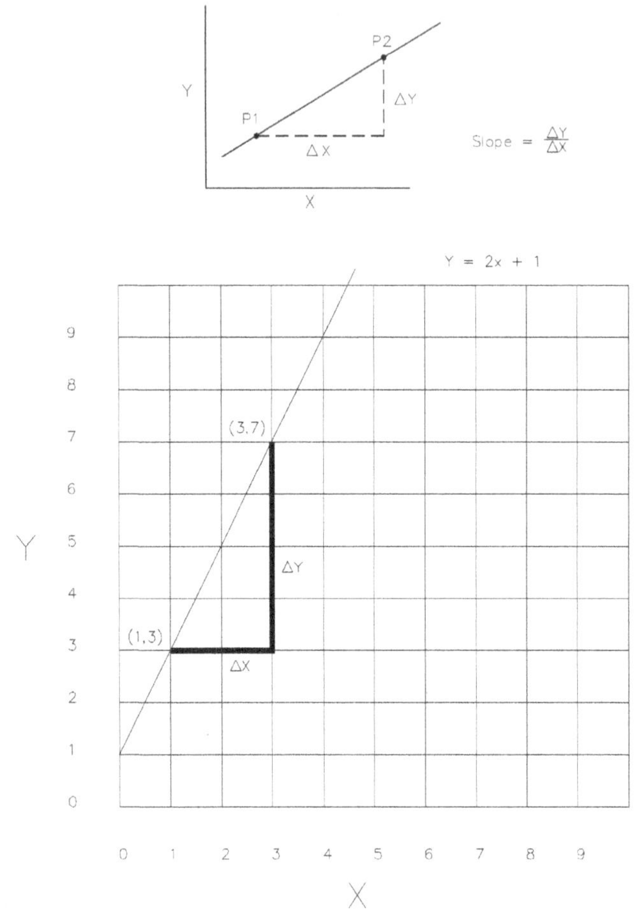

Figure 11 Slope

Since slope m is a measure of the steepness of a line, a slope has the following characteristics:

1. A horizontal line has zero slope.

2. A line that rises to the right has positive slope.

3. A line rising to the left has negative slope.

4. A vertical line has undefined slope because the calculation of the slope would involve division by zero. ($\Delta y/\Delta x$ approaches infinity as the slope approaches vertical.)

Example: What is the slope of the line passing through the points (20, 85) and (30, 125)?

Solution: $m = \frac{125 - 85}{30 - 20} = \frac{40}{10} = 4$

Given the coordinates of the y-intercept where the line crosses the y-axis [written $(0, y)$] and the equation of the line, determine the slope of the line.

The standard linear equation form is $y = mx + b$. If an equation is given in this standard form, m is the slope and b is the y coordinate for the y-intercept.

Example: Determine the slope of the line whose equation is $y = 2x + 3$ and whose y-intercept is (0,3).

Solution: $y = mx + b$

$y = 2x + 3$

$m = 2$

Example: Determine the slope of the line whose equation is $2x + 3y = 6$ and whose y-intercept is (0,2).

Solution: $y = mx + b$

$2x + 3y = 6$ Write in standard form.

$3y = 6 - 2x$

$3y = -2x + 6$

$$y = \frac{-2x + 6}{3}$$

$y = -2/3x + 2$

$m = -2/3$

Example:

Plot the graph of the following linear function. Determine the x-intercept, the y-intercept, and the slope.

$7x + 3y = 21$

Solution: $y = mx + b$

$y = (-7/3)x + 7$

x-intercept = 3
y-intercept = 7

Slope = -2.333

Summary

The important information in this chapter is summarized below.

Slopes Summary

For a straight line, slope is equal to rise over run, or

$$\text{Slope} = \frac{\text{Rise}}{\text{Run}} = \frac{\text{Change in } y}{\text{Change in } x} = \frac{\Delta y}{\Delta x}$$

Since slope m is a measure of the steepness of a line, a slope has the following characteristics:

1. A horizontal line has zero slope.
2. A line that rises to the right of vertical has positive slope.
3. A line rising to the left of vertical has negative slope.
4. A vertical line has undefined slope because the calculation of the slope would involve division by zero ($\Delta y/\Delta x$ approaches infinity as the slope approaches vertical).

INTERPOLATION AND EXTRAPOLATION

This chapter covers the use of interpolation and extrapolation to solve for unknowns on various types of graphs.

EO 1.17 **Given a graph, SOLVE for the unknown using extrapolation.**

EO 1.18 **Given a graph, SOLVE for the unknown using interpolation.**

Definitions

Interpolation: Interpolation is the process of obtaining a value from a graph or table that is located between major points given, or between data points plotted. A ratio process is usually used to obtain the value.

Extrapolation: Extrapolation is the process of obtaining a value from a chart or graph that extends beyond the given data. The "trend" of the data is extended past the last point given and an estimate made of the value.

Interpolation and Extrapolation

Developing a curve from a set of data provides the student with the opportunity to interpolate between given data points. Using the curve in the following example, the value of the dependent variable at 4.5 can be estimated by interpolating on the curve between the two data points given, resulting in the value of 32. Note that the interpolation is the process of obtaining a value on the plotted graph that lies between two given data points. Extrapolation is the process in which information is gained from plotted data by extending the data curve beyond the points of given data (using the basic shape of the curve as a guide), and then estimating the value of a given point by using the extended (extrapolated) curve as the source. The above principles are illustrated in the example that follows.

Example:

Given equation $y = x^2 + 2x + 3$:

Plot the curve for x from 0 to 5.

Extrapolate the curve and give the value of y at $x = 6$.

Put 6 into the equation evaluating y, then compare the values.

Interpolate the curve at $x = 4.5$.

Put 4.5 into the equation evaluating y, then compare the values.

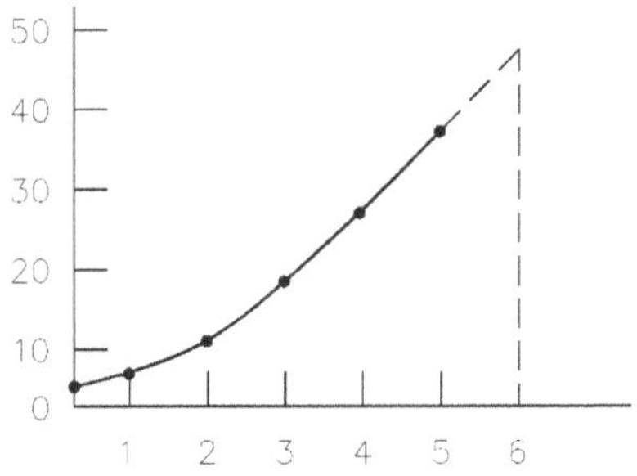

Extrapolating $x = 6$ gives a value of $y = 48$.

Using the equation, the actual value of y is 51.

Interpolating $x = 4.5$ gives a value of $y = 32$.

Using the equation, the actual value of y is 32.25.

Summary

The important information in this chapter is summarized below.

Interpolation and Extrapolation Summary

Interpolation	Interpolation is the process of obtaining a value from a graph or table that is located between major points given, or between data points plotted. A ratio process is usually used to obtain the value.
Extrapolation	Extrapolation is the process of obtaining a value from a chart or graph that extends beyond the given data. The "trend" of the data is extended past the last point given and an estimate made of the value.

DOE-HDBK-1014/2-92
JUNE 1992

DOE FUNDAMENTALS HANDBOOK

MATHEMATICS

Volume 2 of 2

U.S. Department of Energy
Washington, D.C. 20585

FSC-6910

This document has been reproduced directly from the best available copy.

Available to DOE and DOE contractors from the Office of Scientific and Technical Information. P. O. Box 62, Oak Ridge, TN 37831; (615) 576-8401.

Available to the public from the National Technical Information Service, U.S. Department of Commerce, 5285 Port Royal Rd., Springfield, VA 22161.

Order No. DE92019795

ABSTRACT

The *Mathematics Fundamentals Handbook* was developed to assist nuclear facility operating contractors provide operators, maintenance personnel, and the technical staff with the necessary fundamentals training to ensure a basic understanding of mathematics and its application to facility operation. The handbook includes a review of introductory mathematics and the concepts and functional use of algebra, geometry, trigonometry, and calculus. Word problems, equations, calculations, and practical exercises that require the use of each of the mathematical concepts are also presented. This information will provide personnel with a foundation for understanding and performing basic mathematical calculations that are associated with various DOE nuclear facility operations.

Key Words: Training Material, Mathematics, Algebra, Geometry, Trigonometry, Calculus

FOREWORD

The *Department of Energy (DOE) Fundamentals Handbooks* consist of ten academic subjects, which include Mathematics; Classical Physics; Thermodynamics, Heat Transfer, and Fluid Flow; Instrumentation and Control; Electrical Science; Material Science; Mechanical Science; Chemistry; Engineering Symbology, Prints, and Drawings; and Nuclear Physics and Reactor Theory. The handbooks are provided as an aid to DOE nuclear facility contractors.

These handbooks were first published as Reactor Operator Fundamentals Manuals in 1985 for use by DOE category A reactors. The subject areas, subject matter content, and level of detail of the Reactor Operator Fundamentals Manuals were determined from several sources. DOE Category A reactor training managers determined which materials should be included, and served as a primary reference in the initial development phase. Training guidelines from the commercial nuclear power industry, results of job and task analyses, and independent input from contractors and operations-oriented personnel were all considered and included to some degree in developing the text material and learning objectives.

The *DOE Fundamentals Handbooks* represent the needs of various DOE nuclear facilities' fundamental training requirements. To increase their applicability to nonreactor nuclear facilities, the Reactor Operator Fundamentals Manual learning objectives were distributed to the Nuclear Facility Training Coordination Program Steering Committee for review and comment. To update their reactor-specific content, DOE Category A reactor training managers also reviewed and commented on the content. On the basis of feedback from these sources, information that applied to two or more DOE nuclear facilities was considered generic and was included. The final draft of each of the handbooks was then reviewed by these two groups. This approach has resulted in revised modular handbooks that contain sufficient detail such that each facility may adjust the content to fit their specific needs.

Each handbook contains an abstract, a foreword, an overview, learning objectives, and text material, and is divided into modules so that content and order may be modified by individual DOE contractors to suit their specific training needs. Each subject area is supported by a separate examination bank with an answer key.

The *DOE Fundamentals Handbooks* have been prepared for the Assistant Secretary for Nuclear Energy, Office of Nuclear Safety Policy and Standards, by the DOE Training Coordination Program. This program is managed by EG&G Idaho, Inc.

OVERVIEW

The *Department of Energy Fundamentals Handbook* entitled *Mathematics* was prepared as an information resource for personnel who are responsible for the operation of the Department's nuclear facilities. A basic understanding of mathematics is necessary for DOE nuclear facility operators, maintenance personnel, and the technical staff to safely operate and maintain the facility and facility support systems. The information in the handbook is presented to provide a foundation for applying engineering concepts to the job. This knowledge will help personnel more fully understand the impact that their actions may have on the safe and reliable operation of facility components and systems.

The *Mathematics* handbook consists of five modules that are contained in two volumes. The following is a brief description of the information presented in each module of the handbook.

Volume 1 of 2

Module 1 - Review of Introductory Mathematics

This module describes the concepts of addition, subtraction, multiplication, and division involving whole numbers, decimals, fractions, exponents, and radicals. A review of basic calculator operation is included.

Module 2 - Algebra

This module describes the concepts of algebra including quadratic equations and word problems.

Volume 2 of 2

Module 3 - Geometry

This module describes the basic geometric figures of triangles, quadrilaterals, and circles; and the calculation of area and volume.

Module 4 - Trigonometry

This module describes the trigonometric functions of sine, cosine, tangent, cotangent, secant, and cosecant. The use of the pythagorean theorem is also discussed.

Module 5 - Higher Concepts of Mathematics

This module describes logarithmic functions, statistics, complex numbers, imaginary numbers, matrices, and integral and derivative calculus.

The information contained in this handbook is by no means all encompassing. An attempt to present the entire subject of mathematics would be impractical. However, the *Mathematics* handbook does present enough information to provide the reader with a fundamental knowledge level sufficient to understand the advanced theoretical concepts presented in other subject areas, and to better understand basic system and equipment operations.

Department of Energy
Fundamentals Handbook

MATHEMATICS
Module 3
Geometry

www.KnowledgePublications.com

TABLE OF CONTENTS

LIST OF FIGURES

LIST OF TABLES

NONE

REFERENCES

- Dolciani, Mary P., et al., Algebra Structure and Method Book 1, Atlanta: Houghton-Mifflin, 1979.
- Naval Education and Training Command, Mathematics, Vol:1, NAVEDTRA 10069-D1, Washington, D.C.: Naval Education and Training Program Development Center, 1985.
- Olivio, C. Thomas and Olivio, Thomas P., Basic Mathematics Simplified, Albany, NY: Delmar, 1977.
- Science and Fundamental Engineering, Windsor, CT: Combustion Engineering, Inc., 1985.
- Academic Program For Nuclear Power Plant Personnel, Volume 1, Columbia, MD: General Physics Corporation, Library of Congress Card #A 326517, 1982.

TERMINAL OBJECTIVE

1.0 Given a calculator and the correct formula, **APPLY** the laws of geometry to solve mathematical problems.

ENABLING OBJECTIVES

1.1 **IDENTIFY** a given angle as either:
   a. Straight
   b. Acute
   c. Right
   d. Obtuse

1.2 **STATE** the definitions of complimentary and supplementary angles.

1.3 **STATE** the definition of the following types of triangles:
   a. Equilateral
   b. Isosceles
   c. Acute
   d. Obtuse
   e. Scalene

1.4 Given the formula, **CALCULATE** the area and the perimeter of each of the following basic geometric shapes:
   a. Triangle
   b. Parallelogram
   c. Circle

1.5 Given the formula, **CALCULATE** the volume and surface areas of the following solid figures:
   a. Rectangular solid
   b. Cube
   c. Sphere
   d. Right circular cone
   e. Right circular cylinder

Intentionally Left Blank

BASIC CONCEPTS OF GEOMETRY

This chapter covers the basic language and terminology of plane geometry.

EO 1.1 **IDENTIFY a given angle as either:**
- **a. Straight**
- **b. Acute**
- **c. Right**
- **d. Obtuse**

EO 1.2 **STATE the definitions of complimentary and supplementary angles.**

Geometry is one of the oldest branches of mathematics. Applications of geometric constructions were made centuries before the mathematical principles on which the constructions were based were recorded. Geometry is a mathematical study of points, lines, planes, closed flat shapes, and solids. Using any one of these alone, or in combination with others, it is possible to describe, design, and construct every visible object.

The purpose of this section is to provide a foundation of geometric principles and constructions on which many practical problems depend for solution.

Terms

There are a number of terms used in geometry.

1. A *plane* is a flat surface.
2. *Space* is the set of all points.
3. *Surface* is the boundary of a solid.
4. *Solid* is a three-dimensional geometric figure.
5. *Plane geometry* is the geometry of planar figures (two dimensions). Examples are: angles, circles, triangles, and parallelograms.
6. *Solid geometry* is the geometry of three-dimensional figures. Examples are: cubes, cylinders, and spheres.

Lines

A *line* is the path formed by a moving point. A *length of a straight line* is the shortest distance between two nonadjacent points and is made up of collinear points. A *line segment* is a portion of a line. A *ray* is an infinite set of collinear points extending from one end point to infinity. A set of points is noncollinear if the points are not contained in a line.

Two or more straight lines are *parallel* when they are coplanar (contained in the same plane) and do not intersect; that is, when they are an equal distance apart at every point.

Important Facts

The following facts are used frequently in plane geometry. These facts will help you solve problems in this section.

1. The shortest distance between two points is the length of the straight line segment joining them.
2. A straight line segment can be extended indefinitely in both directions.
3. Only one straight line segment can be drawn between two points.
4. A geometric figure can be moved in the plane without any effect on its size or shape.
5. Two straight lines in the same plane are either parallel or they intersect.
6. Two lines parallel to a third line are parallel to each other.

Angles

An *angle* is the union of two nonparallel rays originating from the same point; this point is known as the vertex. The rays are known as sides of the angle, as shown in Figure 1.

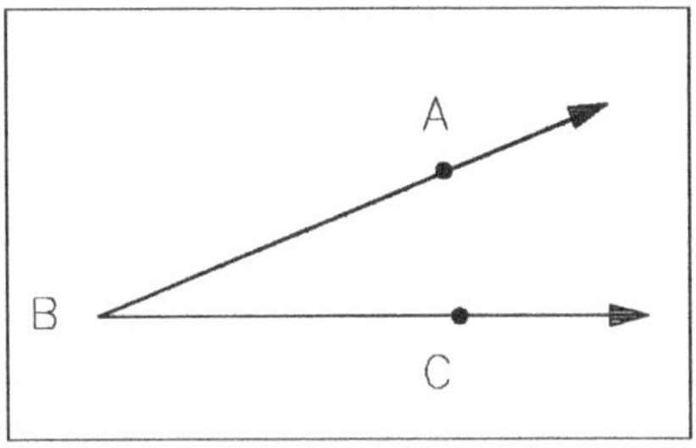

Figure 1 Angle

If ray *AB* is on top of ray *BC*, then the angle *ABC* is a *zero angle*. One complete revolution of a ray gives an angle of 360°.

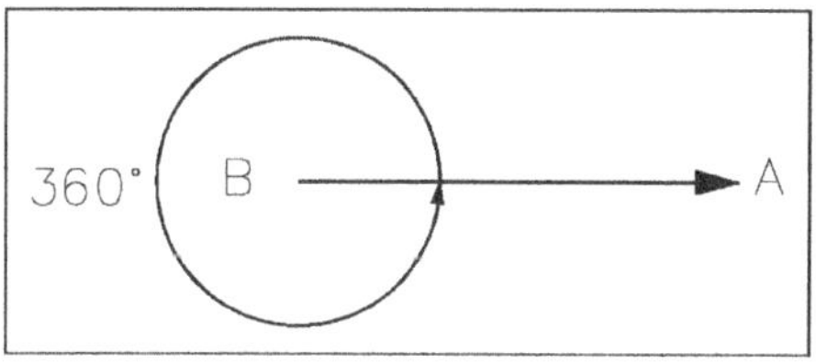

Figure 2 - 360° Angle

Depending on the rotation of a ray, an angle can be classified as right, straight, acute, obtuse, or reflex. These angles are defined as follows:

Right Angle - angle with a ray separated by 90°.

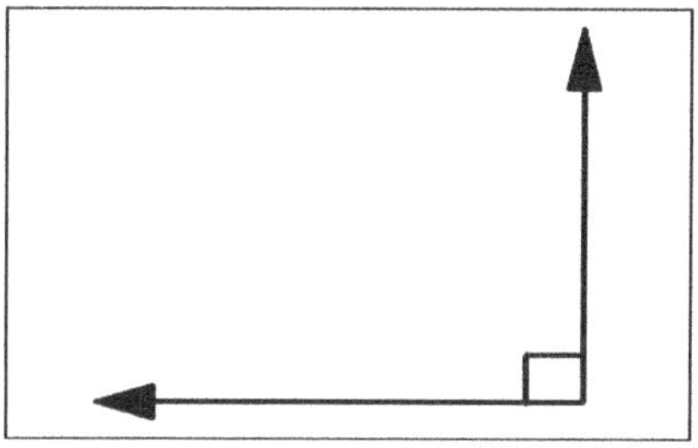

Figure 3 Right Angle

Straight Angle - angle with a ray separated by 180° to form a straight line.

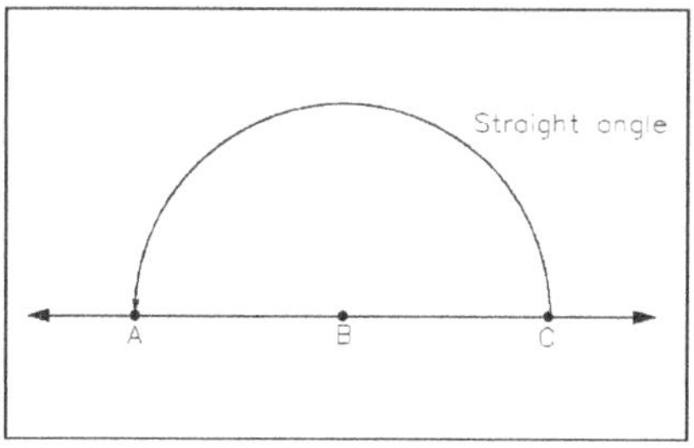

Figure 4 Straight Angle

Acute Angle - angle with a ray separated by less than 90°.

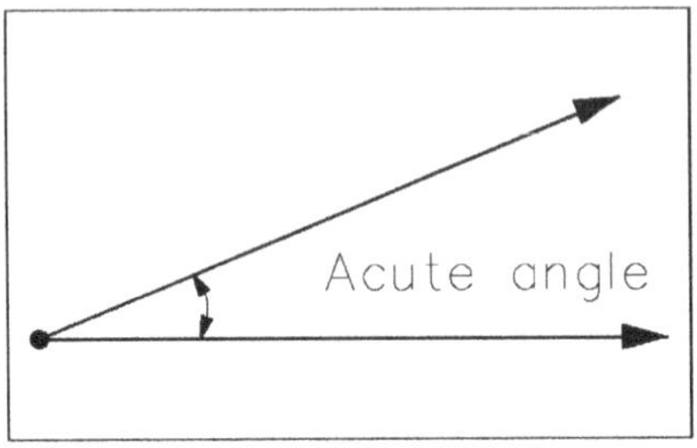

Figure 5 Acute Angle

Obtuse Angle - angle with a ray rotated greater than 90° but less than 180°.

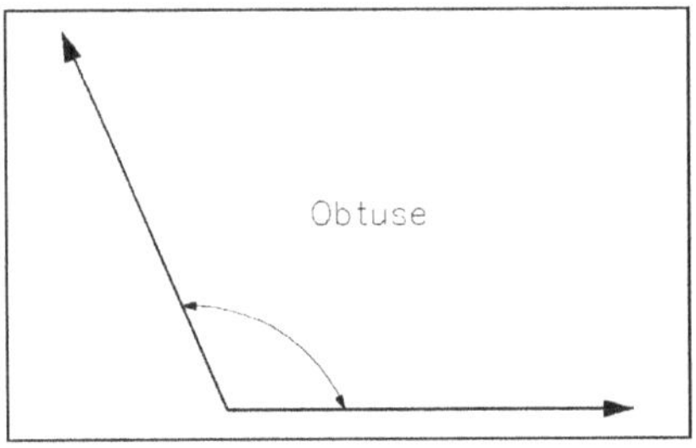

Figure 6 Obtuse Angle

Reflex Angle - angle with a ray rotated greater than 180°.

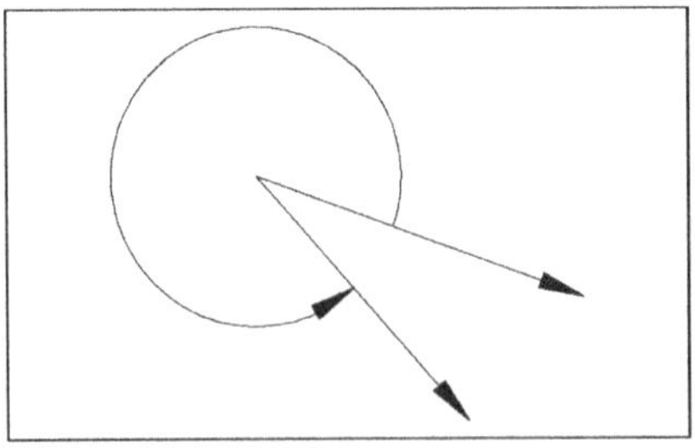

Figure 7 Reflex Angle

If angles are next to each other, they are called *adjacent angles.* If the sum of two angles equals 90°, they are called *complimentary angles*. For example, 27° and 63° are complimentary angles. If the sum of two angles equals 180°, they are called *supplementary angles*. For example, 73° and 107° are supplementary angles.

Summary

The important information in this chapter is summarized below.

Lines and Angles Summary

- Straight lines are parallel when they are in the same plane and do not intersect.
- A straight angle is 180°.
- An acute angle is less than 90°.
- A right angle is 90°.
- An obtuse angle is greater than 90° but less than 180°.
- If the sum of two angles equals 90°, they are complimentary angles.
- If the sum of two angles equals 180°, they are supplementary angles.

SHAPES AND FIGURES OF PLANE GEOMETRY

This chapter covers the calculation of the perimeter and area of selected plane figures.

EO 1.3 STATE the definition of the following types of triangles:
- **a. Equilateral**
- **b. Isosceles**
- **c. Acute**
- **d. Obtuse**
- **e. Scalene**

EO 1.4 Given the formula, CALCULATE the area and the perimeter of each of the following basic geometric shapes:
- **a. Triangle**
- **b. Parallelogram**
- **c. Circle**

The terms and properties of lines, angles, and circles may be applied in the layout, design, development, and construction of closed flat shapes. A new term, plane, must be understood in order to accurately visualize a closed, flat shape. A plane refers to a flat surface on which lies a straight line connecting any two points.

A plane figure is one which can be drawn on a plane surface. There are many types of plane figures encountered in practical problems. Fundamental to most design and construction are three flat shapes: the triangle, the rectangle, and the circle.

Triangles

A *triangle* is a figure formed by using straight line segments to connect three points that are not in a straight line. The straight line segments are called sides of the triangle.

Examples of a number of types of triangles are shown in Figure 8. An *equilateral triangle* is one in which all three sides and all three angles are equal. Triangle *ABC* in Figure 8 is an example of an equilateral triangle. An *isosceles triangle* has two equal sides and two equal angles (triangle *DEF*). A *right triangle* has one of its angles equal to 90° and is the most important triangle for our studies (triangle *GHI*). An *acute triangle* has each of its angles less than 90° (triangle *JKL*). Triangle *MNP* is called a *scalene triangle* because each side is a different length. Triangle *QRS* is considered an *obtuse triangle* since it has one angle greater than 90°. A triangle may have more than one of these attributes. The sum of the interior angles in a triangle is always 180°.

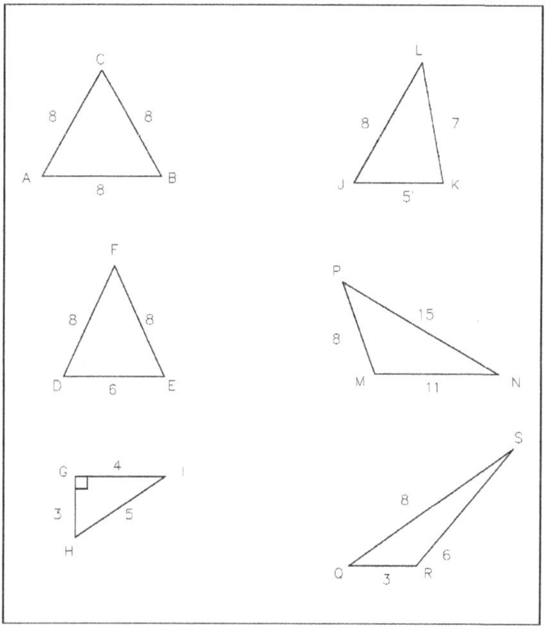

Figure 8 Types of Triangles

Area and Perimeter of Triangles

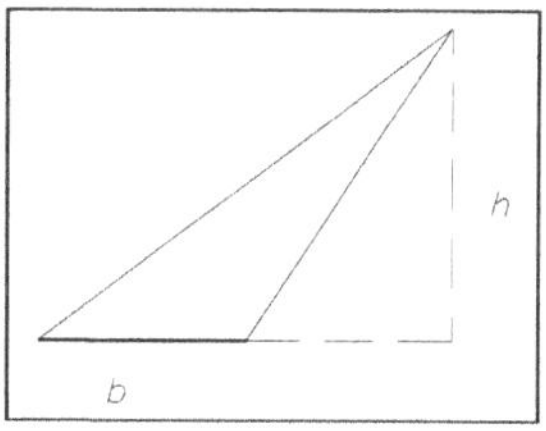

Figure 9 Area of a Triangle

The area of a triangle is calculated using the formula:

$$A = (1/2)(base) \bullet (height) \tag{3-1}$$

or

$$A = (1/2)bh$$

The perimeter of a triangle is calculated using the formula:

$$P = side_1 + side_2 + side_3. \tag{3-2}$$

The area of a traingle is always expressed in square units, and the perimeter of a triangle is always expressed in the original units.

Example:

Calculate the area and perimeter of a right triangle with a 9" base and sides measuring 12" and 15". Be sure to include the units in your answer.

Solution:

$A = 1/2\ bh$
$A = .5(9)(12)$
$A = .5(108)$
$A = 54$ square inches

$P = s_1 + s_2 + b$
$P = 9 + 12 + 15$
$P = 36$ inches

Quadrilaterals

A *quadrilateral* is any four-sided geometric figure.

A *parallelogram* is a four-sided quadrilateral with both pairs of opposite sides parallel, as shown in Figure 10.

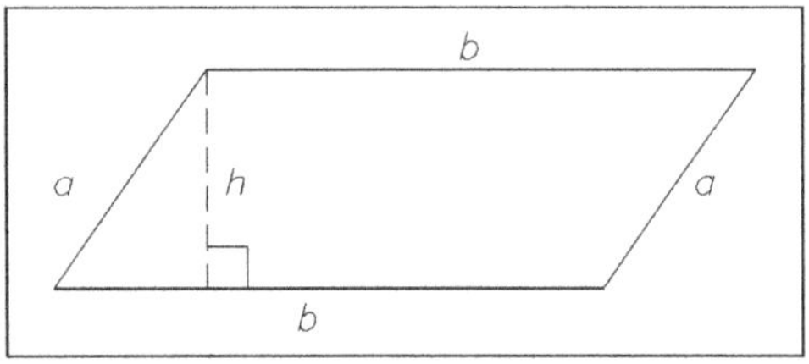

Figure 10 Parallelogram

The area of the parallelogram is calculated using the following formula:

$$A = (base) \bullet (height) = bh \tag{3-3}$$

The perimeter of a parallelogram is calculated using the following formula:

$$P = 2a + 2b \tag{3-4}$$

The area of a parallelogram is always expressed in square units, and the perimeter of a parallelogram is always expressed in the original units.

Example:

Calculate the area and perimeter of a parallelogram with base *(b)* = 4′, height *(h)* = 3′, *a* = 5′ and *b* = 4′. Be sure to include units in your answer.

Solution:

$A = bh$	$P = 2a + 2b$
$A = (4)(3)$	$P = 2(5) + 2(4)$
$A = 12$ square feet	$P = 10 + 8$
	$P = 18$ feet

A *rectangle* is a parallelogram with four right angles, as shown in Figure 11.

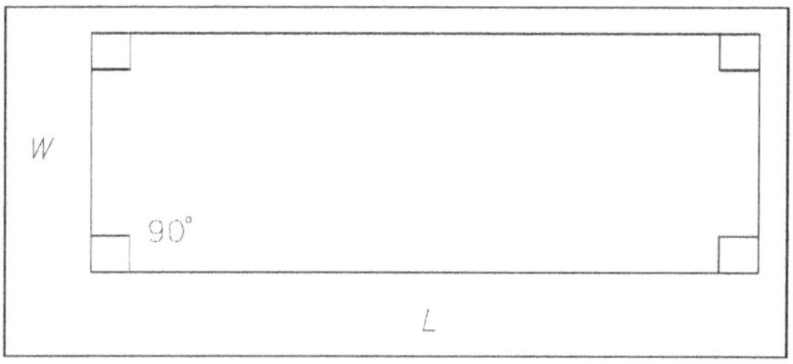

Figure 11 Rectangle

The area of a rectangle is calculated using the following formula:

$$A = (length) \bullet (width) = lw \quad (3\text{-}5)$$

The perimeter of a rectangle is calculated using the following formula:

$$P = 2(length) + 2(width) = 2l + 2w \quad (3\text{-}6)$$

The area of a rectangle is always expressed in square units, and the perimeter of a rectangle is always expressed in the original units.

Example:

Calculate the area and perimeter of a rectangle with $w = 5'$ and $l = 6'$. Be sure to include units in your answer.

Solution:

$A = lw$	$P = 2l + 2w$
$A = (5)(6)$	$P = 2(5) + 2(6)$
$A = 30$ square feet	$P = 10 + 12$
	$P = 22$ feet

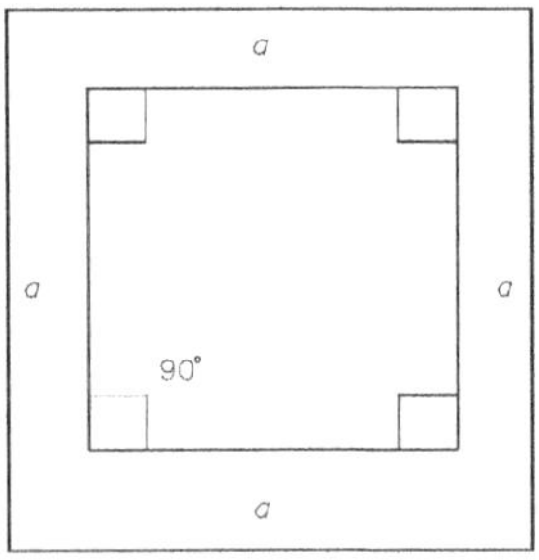

Figure 12 Square

A *square* is a rectangle having four equal sides, as shown in Figure 12.

The area of a square is calculated using the following formula:

$$A = a^2 \tag{3-7}$$

The perimeter of a square is calculated using the following formula:

$$A = 4a \tag{3-8}$$

The area of a square is always expressed in square units, and the perimeter of a square is always expressed in the original units.

Example:

Calculate the area and perimeter of a square with $a = 5'$. Be sure to include units in your answer.

Solution:

$A = a^2$	$P = 4a$
$A = (5)(5)$	$P = 4(5)$
$A = 25$ square feet	$P = 20$ feet

Circles

A *circle* is a plane curve which is equidistant from the center, as shown in Figure 13. The length of the perimeter of a circle is called the *circumference*. The *radius (r)* of a circle is a line segment that joins the center of a circle with any point on its circumference. The *diameter* (*D*) of a circle is a line segment connecting two points of the circle through the center. The area of a circle is calculated using the following formula:

$$A = \pi r^2 \tag{3-9}$$

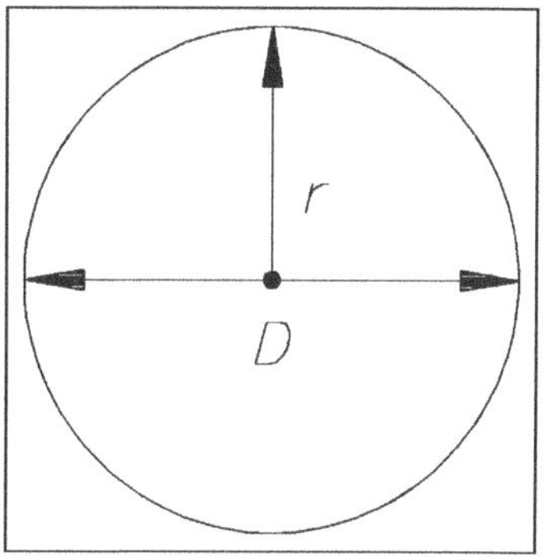

Figure 13 Circle

The circumference of a circle is calculated using the following formula:

$$C = 2\pi r \tag{3-10}$$

or

$$C = \pi D$$

Pi (π) is a theoretical number, approximately 22/7 or 3.141592654, representing the ratio of the circumference to the diameter of a circle. The scientific calculator makes this easy by designating a key for determining π.

The area of a circle is always expressed in square units, and the perimeter of a circle is always expressed in the original units.

Example:

Calculate the area and circumference of a circle with a 3" radius. Be sure to include units in your answer.

Solution:

$A = \pi r^2$	$C = 2\pi r$
$A = \pi(3)(3)$	$C = (2)\pi(3)$
$A = \pi(9)$	$C = \pi(6)$
$A = 28.3$ square inches	$C = 18.9$ inches

Summary

The important information in this chapter is summarized below.

Shapes and Figures of Plane Geometry Summary

• Equilateral Triangle	-	all sides equal
• Isosceles Triangle	-	2 equal sides and 2 equal angles
• Right Triangle	-	1 angle equal to 90°
• Acute Triangle	-	each angle less than 90°
• Obtuse Triangle	-	1 angle greater than 90°
• Scalene Triangle	-	each side a different length
• Area of a triangle	-	$A = (1/2)(base) \bullet (height)$
• Perimeter of a triangle	-	$P = side_1 + side_2 + side_3$
• Area of a parallelogram	-	$A = (base) \bullet (height)$
• Perimeter of a parallelogram	-	$P = 2a + 2b$ where a and b are length of sides
• Area of a rectangle	-	$A = (length) \bullet (width)$
• Perimeter of a rectangle	-	$P = 2(length) + 2(width)$
• Area of a square	-	$A = edge^2$
• Perimeter of a square	-	$P = 4 \text{ x } edge$
• Area of a circle	-	$A = \pi r^2$
• Circumference of a circle	-	$C = 2\pi r$

SOLID GEOMETRIC FIGURES

This chapter covers the calculation of the surface area and volume of selected solid figures.

EO 1.5 Given the formula, CALCULATE the volume and surface areas of the following solid figures:
- **a. Rectangular solid**
- **b. Cube**
- **c. Sphere**
- **d. Right circular cone**
- **e. Right circular cylinder**

The three flat shapes of the triangle, rectangle, and circle may become solids by adding the third dimension of depth. The triangle becomes a cone; the rectangle, a rectangular solid; and the circle, a cylinder.

Rectangular Solids

A *rectangular solid* is a six-sided solid figure with faces that are rectangles, as shown in Figure 14.

The volume of a rectangular solid is calculated using the following formula:

$$V = abc \tag{3-11}$$

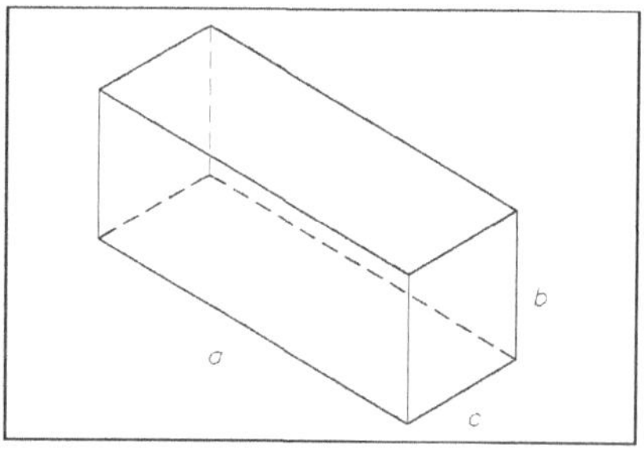

Figure 14 Rectangular Solid

The surface area of a rectangular solid is calculated using the following formula:

$$SA = 2(ab + ac + bc) \tag{3-12}$$

The surface area of a rectangular solid is expressed in square units, and the volume of a rectangular solid is expressed in cubic units.

Example:

Calculate the volume and surface area of a rectangular solid with $a = 3''$, $b = 4''$, and $c = 5''$. Be sure to include units in your answer.

Solution:

$V = (a)(b)(c)$
$V = (3)(4)(5)$
$V = (12)(5)$
$V = 60$ cubic inches

$SA = 2(ab + ac + bc)$
$SA = 2[(3)(4) + (3)(5) + (4)(5)]$
$SA = 2[12 + 15 + 20]$
$SA = 2[47]$
$SA = 94$ square inches

Cube

A *cube* is a six-sided solid figure whose faces are congruent squares, as shown in Figure 15.

The volume of a cube is calculated using the following formula:

$$V = a^3 \quad (3\text{-}13)$$

The surface area of a cube is calculated using the following formula:

$$SA = 6a^2 \quad (3\text{-}14)$$

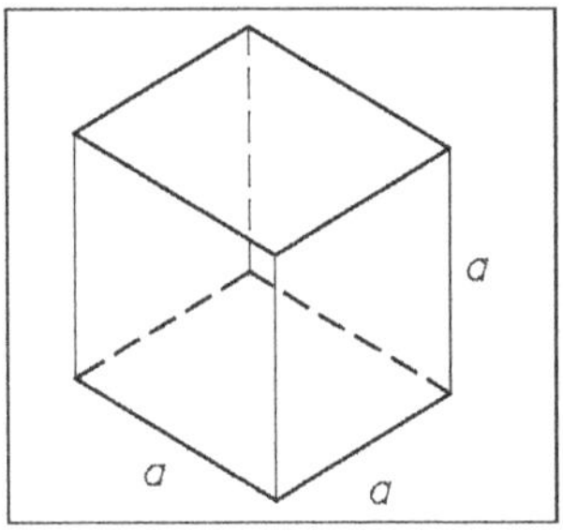

Figure 15 Cube

The surface area of a cube is expressed in square units, and the volume of a cube is expressed in cubic units.

Example:

Calculate the volume and surface area of a cube with $a = 3''$. Be sure to include units in your answer.

Solution:

$V = a^3$
$V = (3)(3)(3)$
$V = 27$ cubic inches

$SA = 6a^2$
$SA = 6(3)(3)$
$SA = 6(9)$
$SA = 54$ square inches

Sphere

A *sphere* is a solid, all points of which are equidistant from a fixed point, the center, as shown in Figure 16.

The volume of a sphere is calculated using the following formula:

$$V = 4/3\pi r^3 \quad (3\text{-}15)$$

The surface area of a sphere is calculated using the following formula:

$$SA = 4\pi r^2 \quad (3\text{-}16)$$

The surface area of a sphere is expressed in square units, and the volume of a sphere is expressed in cubic units.

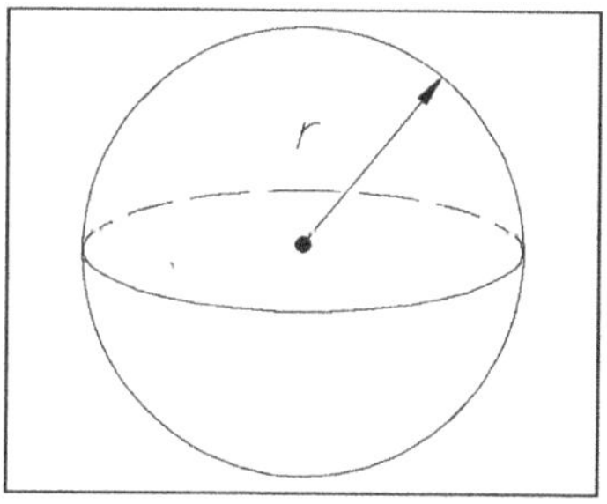

Figure 16 Sphere

Example:

Calculate the volume and surface area of a sphere with $r =$ 4". Be sure to include units in your answer.

Solution:

$V = 4/3\pi r^3$	$SA = 4\pi r^2$
$V = 4/3\pi(4)(4)(4)$	$SA = 4\pi(4)(4)$
$V = 4.2(64)$	$SA = 12.6(16)$
$V = 268.8$ cubic inches	$SA = 201.6$ square inches

Right Circular Cone

A *right circular cone* is a cone whose axis is a line segment joining the vertex to the midpoint of the circular base, as shown in Figure 17.

The volume of a right circular cone is calculated using the following formula:

$$V = 1/3\pi r^2 h \quad (3\text{-}17)$$

The surface area of a right circular cone is calculated using the following formula:

$$SA = \pi r^2 + \pi rl \quad (3\text{-}18)$$

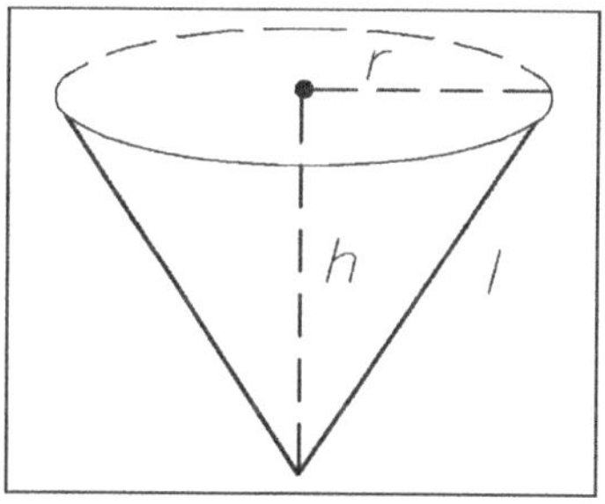

Figure 17 Right Circular Cone

The surface area of a right circular cone is expressed in square units, and the volume of a right circular cone is expressed in cubic units.

Example:

Calculate the volume and surface area of a right circular cone with r = 3", h = 4", and l = 5". Be sure to include the units in your answer.

Solution:

$V = 1/3\pi r^2h$
$V = 1/3\pi(3)(3)(4)$
$V = 1.05(36)$
$V = 37.8$ cubic inches

$SA = \pi r^2 + \pi rl$
$SA = \pi(3)(3) + \pi(3)(5)$
$SA = \pi(9) + \pi(15)$
$SA = 28.3 + 47.1$
$SA = 528/7 = 75\text{-}3/7$ square inches

Right Circular Cylinder

A *right circular cylinder* is a cylinder whose base is perpendicular to its sides. Facility equipment, such as the reactor vessel, oil storage tanks, and water storage tanks, is often of this type.

The volume of a right circular cylinder is calculated using the following formula:

$$V = \pi r^2h \quad (3\text{-}19)$$

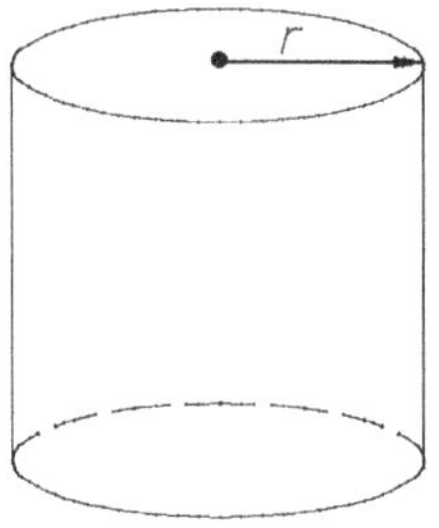

Figure 18 Right Circular Cylinder

The surface area of a right circular cylinder is calculated using the following formula:

$$SA = 2\pi rh + 2\pi r^2 \quad (3\text{-}20)$$

The surface area of a right circular cylinder is expressed in square units, and the volume of a right circular cylinder is expressed in cubic units.

Example:

Calculate the volume and surface area of a right circular cylinder with r = 3" and h = 4". Be sure to include units in your answer.

Solution:

$V = \pi r^2h$
$V = \pi(3)(3)(4)$
$V = \pi(36)$
$V = 113.1$ cubic inches

$SA = 2\pi rh + 2\pi r^2$
$SA = 2\pi(3)(4) + 2\pi(3)(3)$
$SA = 2\pi(12) + 2\pi(9)$
$SA = 132$ square inches

Summary

The important information in this chapter is summarized below.

Solid Geometric Shapes Summary

- Volume of a rectangular solid: abc

  Surface area of a rectangular solid: $2(ab + ac + bc)$
- Volume of a cube: a^3
- Surface area of a cube: $6a^2$
- Volume of a sphere: $4/3\pi r^3$
- Surface area of a sphere: $4\pi r^2$
- Volume of a right circular cone: $1/3\pi r^2h$
- Surface area of a right circular cone: $\pi r^2 + \pi rl$
- Volume of a right circular cylinder: πr^2h
- Surface area of right circular cylinder: $2\pi rh + 2\pi r^2$

Intentionally Left Blank

Department of Energy
Fundamentals Handbook

MATHEMATICS
Module 4
Trigonometry

TABLE OF CONTENTS

LIST OF FIGURES

LIST OF TABLES

REFERENCES

- Academic Program For Nuclear Power Plant Personnel, Volume 1, Columbia, MD: General Physics Corporation, Library of Congress Card #A 326517, 1982.

- Drooyan, I. and Wooton, W., Elementary Algebra and College Students, 6th Edition, John Wiley & Sons, 1984.

- Ellis, R. and Gulick, D., College Algebra and Trigonometry, 2nd Edition, Harcourt Brace Jouanovich, Publishers, 1984.

- Rice, B.J. and Strange, J.D., Plane Trigonometry, 2nd Edition, Prinole, Weber & Schmidt, Inc., 1978.

TERMINAL OBJECTIVE

1.0 Given a calculator and a list of formulas, **APPLY** the laws of trigonometry to solve for unknown values.

ENABLING OBJECTIVES

1.1 Given a problem, **APPLY** the Pythagorean theorem to solve for the unknown values of a right triangle.

1.2 Given the following trigonometric terms, **IDENTIFY** the related function:

a. Sine
b. Cosine
c. Tangent
d. Cotangent
e. Secant
f. Cosecant

1.3 Given a problem, **APPLY** the trigonometric functions to solve for the unknown.

1.4 **STATE** the definition of a radian.

Intentionally Left Blank

PYTHAGOREAN THEOREM

This chapter covers right triangles and solving for unknowns using the Pythagorean theorem.

EO 1.1 Given a problem, APPLY the Pythagorean theorem to solve for the unknown values of a right triangle.

Trigonometry is the branch of mathematics that is the study of angles and the relationship between angles and the lines that form them. Trigonometry is used in Classical Physics and Electrical Science to analyze many physical phenomena. Engineers and operators use this branch of mathematics to solve problems encountered in the classroom and on the job. The most important application of trigonometry is the solution of problems involving triangles, particularly right triangles.

Trigonometry is one of the most useful branches of mathematics. It is used to indirectly measure distances which are difficult to measure directly. For example, the height of a flagpole or the distance across a river can be measured using trigonometry.

As shown in Figure 1 below, a triangle is a plane figure formed using straight line segments (*AB*, *BC*, *CA*) to connect three points (*A*, *B*, *C*) that are not in a straight line. The sum of the measures of the three interior angles (a', b', c') is 180°, and the sum of the lengths of any two sides is always greater than or equal to the third.

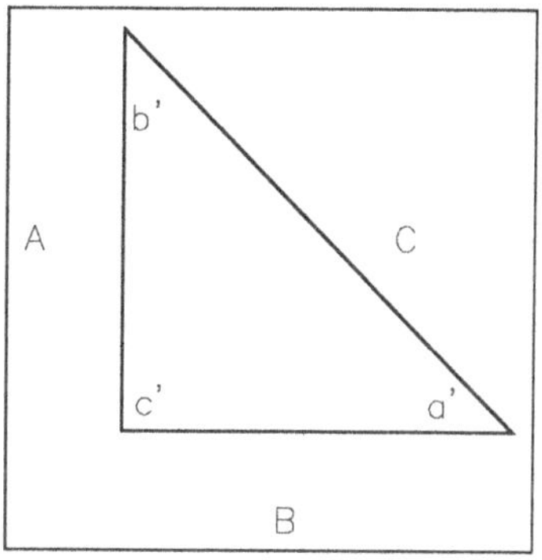

Figure 1 Triangle

Pythagorean Theorem

The Pythagorean theorem is a tool that can be used to solve for unknown values on right triangles. In order to use the Pythagorean theorem, a term must be defined. The term *hypotenuse* is used to describe the side of a right triangle opposite the right angle. Line segment C is the hypotenuse of the triangle in Figure 1.

The Pythagorean theorem states that in any right triangle, the square of the length of the hypotenuse equals the sum of the squares of the lengths of the other two sides.

This may be written as $c^2 = a^2 + b^2$ or $c = \sqrt{a^2 + b^2}$. (4-1)

Example:

The two legs of a right triangle are 5 ft and 12 ft. How long is the hypotenuse?

Let the hypotenuse be c ft.

$$a^2 + b^2 = c^2$$

$$12^2 + 5^2 = c^2$$

$$144 + 25 = c^2$$

$$169 = c^2$$

$$\sqrt{169} = c$$

$$13 \text{ ft} = c$$

Using the Pythagorean theorem, one can determine the value of the unknown side of a right triangle when given the value of the other two sides.

Example:

Given that the hypotenuse of a right triangle is 18" and the length of one side is 11", what is the length of the other side?

$$a^2 + b^2 = c^2$$

$$11^2 + b^2 = 18^2$$

$$b^2 = 18^2 - 11^2$$

$$b^2 = 324 - 121$$

$$b = \sqrt{203}$$

$$b = 14.2 \text{ in}$$

Summary

The important information in this chapter is summarized below.

Pythagorean Theorem Summary

- The Pythagorean theorem states that in any right triangle, the square of the length of the hypotenuse equals the sum of the squares of the lengths of the other two sides.

  This may be written as $c^2 = a^2 + b^2$ or $c = \sqrt{a^2 + b^2}$.

TRIGONOMETRIC FUNCTIONS

This chapter covers the six trigonometric functions and solving right triangles.

EO 1.2 Given the following trigonometric terms, IDENTIFY the related function:

a. Sine
b. Cosine
c. Tangent
d. Cotangent
e. Secant
f. Cosecant

EO 1.3 Given a problem, APPLY the trigonometric functions to solve for the unknown.

As shown in the previous chapter, the lengths of the sides of right triangles can be solved using the Pythagorean theorem. We learned that if the lengths of two sides are known, the length of the third side can then be determined using the Pythagorean theorem. One fact about triangles is that the sum of the three angles equals 180°. If right triangles have one 90° angle, then the sum of the other two angles must equal 90°. Understanding this, we can solve for the unknown angles if we know the length of two sides of a right triangle. This can be done by using the six trigonometric functions.

In right triangles, the two sides (other than the hypotenuse) are referred to as the opposite and adjacent sides. In Figure 2, side a is the opposite side of the angle θ and side b is the adjacent side of the angle θ. The terms hypotenuse, opposite side, and adjacent side are used to distinguish the relationship between an acute angle of a right triangle and its sides. This relationship is given by the six trigonometric functions listed below:

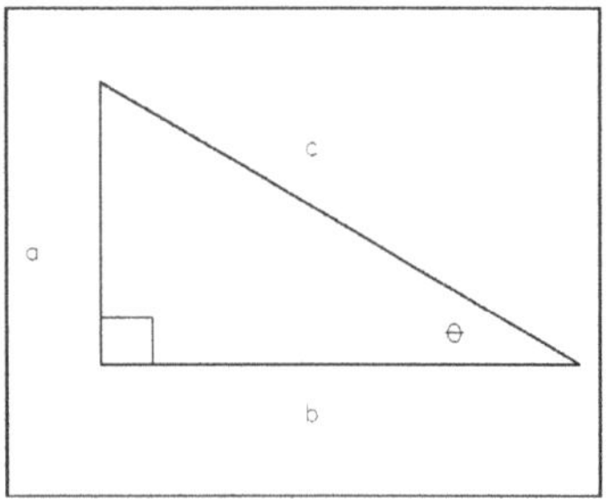

Figure 2 Right Triangle

$$\text{sine } \theta = \frac{a}{c} = \frac{\text{opposite}}{\text{hypotenuse}} \qquad (4\text{-}2)$$

$$\text{cosine } \theta = \frac{b}{c} = \frac{\text{adjacent}}{\text{hypotenuse}} \qquad (4\text{-}3)$$

$$\text{tangent } \theta = \frac{a}{b} = \frac{\text{opposite}}{\text{adjacent}} \tag{4-4}$$

$$\text{cosecant } \theta = \frac{c}{b} = \frac{\text{hypotenuse}}{\text{oposite}} \tag{4-5}$$

$$\text{secant } \theta = \frac{c}{a} = \frac{\text{hypotenuse}}{\text{adjacent}} \tag{4-6}$$

$$\text{cotangent } \theta = \frac{b}{a} = \frac{\text{adjacent}}{\text{opposite}} \tag{4-7}$$

The trigonometric value for any angle can be determined easily with the aid of a calculator. To find the sine, cosine, or tangent of any angle, enter the value of the angle into the calculator and press the desired function. Note that the secant, cosecant, and cotangent are the mathematical inverse of the sine, cosine and tangent, respectively. Therefore, to determine the cotangent, secant, or cosecant, first press the SIN, COS, or TAN key, then press the INV key.

Example:

Determine the values of the six trigonometric functions of an angle formed by the x-axis and a line connecting the origin and the point (3,4).

Solution:

To help to "see" the solution of the problem it helps to plot the points and construct the right triangle.

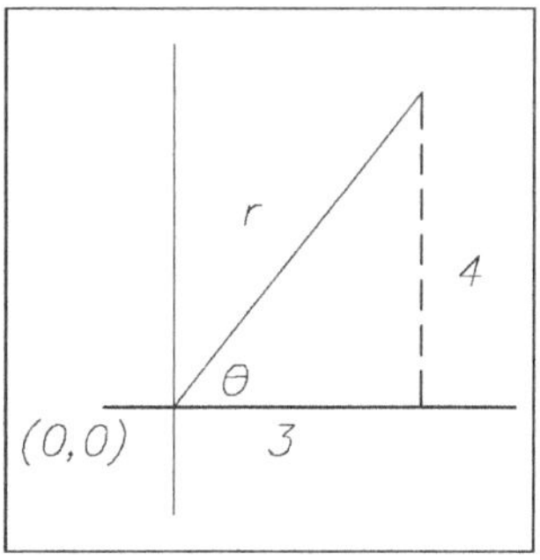

Figure 3 Example Problem

Label all the known angles and sides, as shown in Figure 3.

From the triangle, we can see that two of the sides are known. But to answer the problem, all three sides must be determined. Therefore the Pythagorean theorem must be applied to solve for the unknown side of the triangle.

$$x = 3$$

$$y = 4$$

$$r = \sqrt{x^2 + y^2} = \sqrt{3^2 + 4^2}$$

$$r = \sqrt{9 + 16} = \sqrt{25} = 5$$

Having solved for all three sides of the triangle, the trigonometric functions can now be determined. Substitute the values for *x*, *y*, and *r* into the trigonometric functions and solve.

$$\sin\theta = \frac{y}{r} = \frac{4}{5} = 0.800$$

$$\cos\theta = \frac{x}{r} = \frac{3}{5} = 0.600$$

$$\tan\theta = \frac{y}{x} = \frac{4}{3} = 1.333$$

$$\csc\theta = \frac{r}{y} = \frac{5}{4} = 1.250$$

$$\sec\theta = \frac{r}{x} = \frac{5}{3} = 1.667$$

$$\cot\theta = \frac{x}{y} = \frac{3}{4} = 0.750$$

Although the trigonometric functions of angles are defined in terms of lengths of the sides of right triangles, they are really functions of the angles only. The numerical values of the trigonometric functions of any angle depend on the size of the angle and not on the length of the sides of the angle. Thus, the sine of a 30° angle is always 1/2 or 0.500.

Inverse Trigonometric Functions

When the value of a trigonometric function of an angle is known, the size of the angle can be found. The inverse trigonometric function, also known as the arc function, defines the angle based on the value of the trigonometric function. For example, the sine of 21° equals 0.35837; thus, the arc sine of 0.35837 is 21°.

There are two notations commonly used to indicate an inverse trigonometric function.

$$\arcsin 0.35837 = 21°$$

$$\sin^{-1} 0.35837 = 21°$$

The notation *arcsin* means *the angle whose sine is.* The notation *arc* can be used as a prefix to any of the trigonometric functions. Similarly, the notation sin^{-1} means *the angle whose sine is.* It is important to remember that the -1 in this notation is not a negative exponent but merely an indication of the inverse trigonometric function.

To perform this function on a calculator, enter the numerical value, press the INV key, then the SIN, COS, or TAN key. To calculate the inverse function of cot, csc, and sec, the reciprocal key must be pressed first then the SIN, COS, or TAN key.

Examples:

Evaluate the following inverse trigonometric functions.

$$\arcsin 0.3746 = 22°$$

$$\arccos 0.3746 = 69°$$

$$\arctan 0.3839 = 21°$$

$$\text{arccot } 2.1445 = \arctan \frac{1}{2.1445} = \arctan 0.4663 = 25°$$

$$\text{arcsec } 2.6695 = \arccos \frac{1}{2.6695} = \arccos 0.3746 = 68°$$

$$\text{arccsc } 2.7904 = \arcsin \frac{1}{2.7904} = \arcsin 0.3584 = 21°$$

Summary

The important information in this chapter is summarized below.

Trigonometric Functions Summary

- The six trigonometric functions are:

$$\text{sine } \theta = \frac{a}{c} = \frac{\text{opposite}}{\text{hypotenuse}}$$

$$\text{cosine } \theta = \frac{b}{c} = \frac{\text{adjacent}}{\text{hypotenuse}}$$

$$\text{tangent } \theta = \frac{a}{b} = \frac{\text{opposite}}{\text{adjacent}}$$

$$\text{cotangent } \theta = \frac{b}{a} = \frac{\text{adjacent}}{\text{opposite}}$$

$$\text{cosecant } \theta = \frac{c}{b} = \frac{\text{hypotenuse}}{\text{opposite}}$$

$$\text{secant } \theta = \frac{c}{a} = \frac{\text{hypotenuse}}{\text{adjacent}}$$

RADIANS

This chapter will cover the measure of angles in terms of radians and degrees.

EO 1.4 STATE the definition of a radian.

Radian Measure

The size of an angle is usually measured in degrees. However, in some applications the size of an angle is measured in radians. A radian is defined in terms of the length of an arc subtended by an angle at the center of a circle. An angle whose size is one radian subtends an arc whose length equals the radius of the circle. Figure 4 shows $\angle BAC$ whose size is one radian. The length of arc BC equals the radius r of the circle. The size of an angle, in radians, equals the length of the arc it subtends divided by the radius.

$$\text{Radians} = \frac{\text{Length of Arc}}{\text{Radius}} \tag{4-8}$$

One radian equals approximately 57.3 degrees. There are exactly 2π radians in a complete revolution. Thus 2π radians equals 360 degrees: π radians equals 180 degrees.

Although the radian is defined in terms of the length of an arc, it can be used to measure any angle. Radian measure and degree measure can be converted directly. The size of an angle in degrees is changed to radians by multiplying by $\frac{\pi}{180}$. The size of an angle in radians is changed to degrees by multiplying by $\frac{180}{\pi}$.

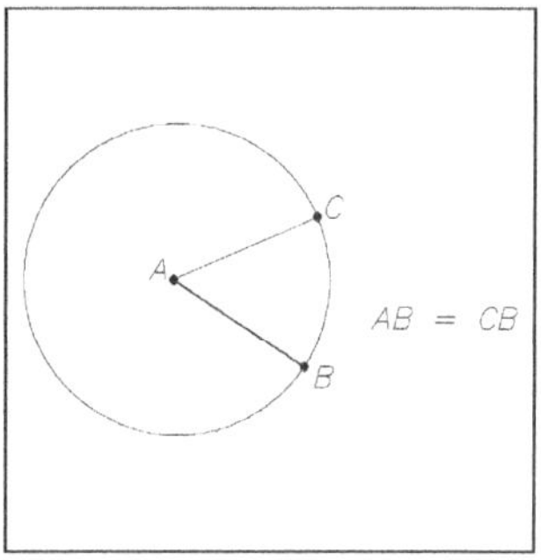

Figure 4 Radian Angle

Example:

Change 68.6° to radians.

$$068.6°\left(\frac{\pi}{180}\right) = \frac{(68.6)\pi}{180} = 1.20 \text{ radians}$$

Example:

Change 1.508 radians to degrees.

$$(1.508 \text{ radians})\left(\frac{180}{\pi}\right) = \frac{(1.508)(180)}{\pi} = 86.4°$$

Summary

The important information in this chapter is summarized below.

Radian Measure Summary

- A radian equals approximately 57.3° and is defined as the angle subtended by an arc whose length is equal to the radius of the circle.

  $$\text{Radian} = \frac{\text{Length of arc}}{\text{Radius of circle}}$$

  π radians = 180°

Department of Energy
Fundamentals Handbook

MATHEMATICS
Module 5
Higher Concepts of Mathematics

TABLE OF CONTENTS

TABLE OF CONTENTS (Cont)

LIST OF FIGURES

LIST OF TABLES

NONE

REFERENCES

- Dolciani, Mary P., et al., Algebra Structure and Method Book 1, Atlanta: Houghton-Mifflin, 1979.
- Naval Education and Training Command, Mathematics, Vol:3, NAVEDTRA 10073-A, Washington, D.C.: Naval Education and Training Program Development Center, 1969.
- Olivio, C. Thomas and Olivio, Thomas P., Basic Mathematics Simplified, Albany, NY: Delmar, 1977.
- Science and Fundamental Engineering, Windsor, CT: Combustion Engineering, Inc., 1985.
- Academic Program For Nuclear Power Plant Personnel, Volume 1, Columbia, MD: General Physics Corporation, Library of Congress Card #A 326517, 1982.
- Standard Mathematical Tables, 23rd Edition, Cleveland, OH: CRC Press, Inc., Library of Congress Card #30-4052, ISBN 0-87819-622-6, 1975.

TERMINAL OBJECTIVE

1.0 **SOLVE** problems involving probability and simple statistics.

ENABLING OBJECTIVES

1.1 **STATE** the definition of the following statistical terms:
- a. Mean
- b. Variance
- c. Mean variance

1.2 **CALCULATE** the mathematical mean of a given set of data.

1.3 **CALCULATE** the mathematical mean variance of a given set of data.

1.4 Given the data, **CALCULATE** the probability of an event.

TERMINAL OBJECTIVE

2.0 **SOLVE** for problems involving the use of complex numbers.

ENABLING OBJECTIVES

2.1 **STATE** the definition of an imaginary number.

2.2 **STATE** the definition of a complex number.

2.3 **APPLY** the arithmetic operations of addition, subtraction, multiplication, and division to complex numbers.

TERMINAL OBJECTIVE

3.0 **SOLVE** for the unknowns in a problem through the application of matrix mathematics.

ENABLING OBJECTIVES

3.1 **DETERMINE** the dimensions of a given matrix.

3.2 **SOLVE** a given set of equations using Cramer's Rule.

TERMINAL OBJECTIVE

4.0 **DESCRIBE** the use of differentials and integration in mathematical problems.

ENABLING OBJECTIVES

4.1 **STATE** the graphical definition of a derivative.

4.2 **STATE** the graphical definition of an integral.

Intentionally Left Blank

STATISTICS

This chapter will cover the basic concepts of statistics.

EO 1.1 STATE the definition of the following statistical terms:
a. Mean
b. Variance
c. Mean variance

EO 1.2 CALCULATE the mathematical mean of a given set of data.

EO 1.3 CALCULATE the mathematical mean variance of a given set of data.

EO 1.4 Given the data, CALCULATE the probability of an event.

In almost every aspect of an operator's work, there is a necessity for making decisions resulting in some significant action. Many of these decisions are made through past experience with other similar situations. One might say the operator has developed a method of intuitive inference: unconsciously exercising some principles of probability in conjunction with statistical inference following from observation, and arriving at decisions which have a high chance of resulting in expected outcomes. In other words, statistics is a method or technique which will enable us to approach a problem of determining a course of action in a systematic manner in order to reach the desired results.

Mathematically, statistics is the collection of great masses of numerical information that is summarized and then analyzed for the purpose of making decisions; that is, the use of past information is used to predict future actions. In this chapter, we will look at some of the basic concepts and principles of statistics.

Frequency Distribution

When groups of numbers are organized, or ordered by some method, and put into tabular or graphic form, the result will show the "frequency distribution" of the data.

Example:

A test was given and the following grades were received: the number of students receiving each grade is given in parentheses.

99(1), 98(2), 96(4), 92(7), 90(5), 88(13), 86(11), 83(7), 80(5), 78(4), 75(3), 60(1)

The data, as presented, is arranged in descending order and is referred to as an ordered array. But, as given, it is difficult to determine any trend or other information from the data. However, if the data is tabled and/or plotted some additional information may be obtained. When the data is ordered as shown, a frequency distribution can be seen that was not apparent in the previous list of grades.

Grades	Number of Occurrences	Frequency Distribution
99	1	1
98	11	2
96	1111	4
92	11111 11	7
90	11111	5
88	11111 11111 111	13
86	11111 11111 1	11
83	11111 11	7
80	11111	5
78	1111	4
75	111	3
	1	1

In summary, one method of obtaining additional information from a set of data is to determine the frequency distribution of the data. The frequency distribution of any one data point is the number of times that value occurs in a set of data. As will be shown later in this chapter, this will help simplify the calculation of other statistically useful numbers from a given set of data.

The Mean

One of the most common uses of statistics is the determination of the mean value of a set of measurements. The term "Mean" is the statistical word used to state the "average" value of a set of data. The mean is mathematically determined in the same way as the "average" of a group of numbers is determined.

The arithmetic mean of a set of N measurements, $X_1, X_2, X_3, \ldots, X_N$ is equal to the sum of the measurements divided by the number of data points, N. Mathematically, this is expressed by the following equation:

$$\bar{x} = \frac{1}{n}\sum_{i=1}^{n} x_i$$

where

$\bar{x}$	=	the mean
n	=	the number of values (data)
x_1	=	the first data point, x_2 = the second data point,....x_i = the ith data point
x_i	=	the ith data point, x_1 = the first data point, x_2 = the second data point, etc.

The symbol Sigma (Σ) is used to indicate summation, and $i = 1$ to n indicates that the values of x_i from $i = 1$ to $i = n$ are added. The sum is then divided by the number of terms added, n.

Example:

Determine the mean of the following numbers:

5, 7, 1, 3, 4

Solution:

$$\bar{x} = \frac{1}{n}\sum_{i=1}^{n} x_i = \frac{1}{5}\sum_{i=1}^{5} x_i$$

where

$\bar{x}$	=	the mean
n	=	the number of values (data) = 5
x_1	=	5, $x_2 = 7$, $x_3 = 1$, $x_4 = 3$, $x_5 = 4$

substituting

$$\bar{x} = (5 + 7 + 1 + 3 + 4)/5 = 20/5 = 4$$

4 is the mean.

Example:

Find the mean of 67, 88, 91, 83, 79, 81, 69, and 74.

Solution:

$$\bar{x} = \frac{1}{n}\sum_{i=1}^{n} x_i$$

The sum of the scores is 632 and $n = 8$, therefore

$$\bar{x} = \frac{632}{8}$$

$$\bar{x} = 79$$

In many cases involving statistical analysis, literally hundreds or thousands of data points are involved. In such large groups of data, the frequency distribution can be plotted and the calculation of the mean can be simplified by multiplying each data point by its frequency distribution, rather than by summing each value. This is especially true when the number of discrete values is small, but the number of data points is large.

Therefore, in cases where there is a recurring number of data points, like taking the mean of a set of temperature readings, it is easier to multiply each reading by its frequency of occurrence (frequency of distribution), then adding each of the multiple terms to find the mean. This is one application using the frequency distribution values of a given set of data.

Example:

Given the following temperature readings,

573, 573, 574, 574, 574, 574, 575, 575, 575, 575, 575, 576, 576, 576, 578

Solution:

Determine the frequency of each reading.

Frequency Distribution

Temperatures	Frequency (f)	$(f)(x_i)$
573	2	1146
574	4	2296
575	5	2875
576	3	1728
578	1	578
	15	8623

Then calculate the mean,

$$\bar{x} = \frac{1}{n}\sum_{i=1}^{n} x_i$$

$$\bar{x} = \frac{2(573) + 4(574) + 5(575) + 3(576) + 1(578)}{15}$$

$$\bar{x} = \frac{8623}{15}$$

$$\bar{x} = 574.9$$

Variability

We have discussed the averages and the means of sets of values. While the mean is a useful tool in describing a characteristic of a set of numbers, sometimes it is valuable to obtain information about the mean. There is a second number that indicates how representative the mean is of the data. For example, in the group of numbers, 100, 5, 20, 2, the mean is 31.75. If these data points represent tank levels for four days, the use of the mean level, 31.75, to make a decision using tank usage could be misleading because none of the data points was close to the mean.

This spread, or distance, of each data point from the mean is called the *variance.* The variance of each data point is calculated by:

$$\text{Variance} = \bar{x} - x_i$$

where

x_i = each data point

$\bar{x}$ = mean

The variance of each data point does not provide us with any useful information. But if the mean of the variances is calculated, a very useful number is determined. The *mean variance* is the average value of the variances of a set of data. The mean variance is calculated as follows:

$$\text{Mean Variance} = \frac{1}{n}\sum_{i=1}^{n} \left| x_i - \bar{x} \right|$$

The mean variance, or mean deviation, can be calculated and used to make judgments by providing information on the quality of the data. For example, if you were trying to decide whether to buy stock, and all you knew was that this month's average price was $10, and today's price is $9, you might be tempted to buy some. But, if you also knew that the mean variance in the stock's price over the month was $6, you would realize the stock had fluctuated widely during the month. Therefore, the stock represented a more risky purchase than just the average price indicated.

It can be seen that to make sound decisions using statistical data, it is important to analyze the data thoroughly before making any decisions.

Example:

Calculate the variance and mean variance of the following set of hourly tank levels. Assume the tank is a 100 gal. tank. Based on the mean and the mean variance, would you expect the tank to be able to accept a 40% (40 gal.) increase in level at any time?

1:00 - 40%	6:00 - 38%	11:00- 34%
2:00 - 38%	7:00 - 34%	12:00- 30%
3:00 - 28%	8:00 - 28%	1:00 - 40%
4:00 - 28%	9:00 - 40%	2:00 - 36%
5:00 - 40%	10:00- 38%	

Solution:

The mean is

[40(4)+38(3)+36+34(2)+30+28(3)]/14= 492/14 = 35.1

The mean variance is:

$$\frac{1}{14}\left(|40 - 35.1| + |38 - 35.1| + |28 - 35.1| + ...|36 - 35.1|\right) =$$

$$\frac{1}{14}(57.8) = 4.12$$

From the tank mean of 35.1%, it can be seen that a 40% increase in level will statistically fit into the tank; 35.1 + 40 <100%. But, the mean doesn't tell us if the level varies significantly over time. Knowing the mean variance is 4.12% provides the additional information. Knowing the mean variance also allows us to infer that the level at any given time (most likely) will not be greater than 35.1 + 4.12 = 39.1%; and 39.1 + 40 is still less than 100%. Therefore, it is a good assumption that, in the near future, a 40% level increase will be accepted by the tank without any spillage.

Normal Distribution

The concept of a normal distribution curve is used frequently in statistics. In essence, a normal distribution curve results when a large number of random variables are observed in nature, and their values are plotted. While this "distribution" of values may take a variety of shapes, it is interesting to note that a very large number of occurrences observed in nature possess a frequency distribution which is approximately bell-shaped, or in the form of a normal distribution, as indicated in Figure 1.

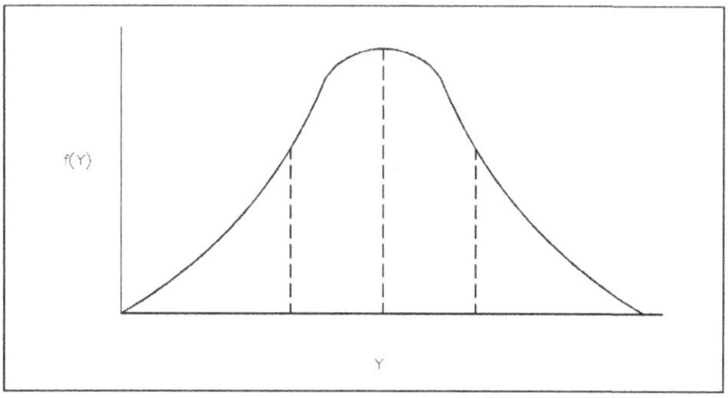

Figure 1 Graph of a Normal Probability Distribution

The significance of a normal distribution existing in a series of measurements is two fold. First, it explains why such measurements tend to possess a normal distribution; and second, it provides a valid basis for statistical inference. Many estimators and decision makers that are used to make inferences about large numbers of data, are really sums or averages of those measurements. When these measurements are taken, especially if a large number of them exist, confidence can be gained in the values, if these values form a bell-shaped curve when plotted on a distribution basis.

Probability

If E_1 is the number of heads, and E_2 is the number of tails, $E_1/(E_1 + E_2)$ is an experimental determination of the probability of heads resulting when a coin is flipped.

$$P(E_1) = n/N$$

By definition, the probability of an event must be greater than or equal to 0, and less than or equal to 1. In addition, the sum of the probabilities of all outcomes over the entire "event" must add to equal 1. For example, the probability of heads in a flip of a coin is 50%, the probability of tails is 50%. If we assume these are the only two possible outcomes, 50% + 50%, the two outcomes, equals 100%, or 1.

The concept of probability is used in statistics when considering the reliability of the data or the measuring device, or in the correctness of a decision. To have confidence in the values measured or decisions made, one must have an assurance that the probability is high of the measurement being true, or the decision being correct.

To calculate the probability of an event, the number of successes (s), and failures (f), must be determined. Once this is determined, the probability of the success can be calculated by:

$$p = \frac{s}{s + f}$$

where

$$s + f = n = \text{number of tries.}$$

Example:

Using a die, what is the probability of rolling a three on the first try?

Solution:

First, determine the number of possible outcomes. In this case, there are 6 possible outcomes. From the stated problem, the roll is a success only if a 3 is rolled. There is only 1 success outcome and 5 failures. Therefore,

Probability = 1/(1+5)
= 1/6

In calculating probability, the probability of a series of independent events equals the product of probability of the individual events.

Example:

Using a die, what is the probability of rolling two 3s in a row?

Solution:

From the previous example, there is a 1/6 chance of rolling a three on a single throw. Therefore, the chance of rolling two threes is:

1/6 x 1/6 = 1/36

one in 36 tries.

Example:

An elementary game is played by rolling a die and drawing a ball from a bag containing 3 white and 7 black balls. The player wins whenever he rolls a number less than 4 and draws a black ball. What is the probability of winning in the first attempt?

Solution:

There are 3 successful outcomes for rolling less than a 4, (i.e. 1,2,3). The probability of rolling a 3 or less is:

3/(3+3) = 3/6 = 1/2 or 50%.

The probability of drawing a black ball is:

7/(7+3) = 7/10.

Therefore, the probability of both events happening at the same time is:

7/10 x 1/2 = 7/20.

Summary

The important information in this chapter is summarized below.

Statistics Summary

Mean	-	The sum of a group of values divided by the number of values.
Frequency Distribution	-	An arrangement of statistical data that exhibits the frequency of the occurrence of the values of a variable.
Variance	-	The difference of a data point from the mean.
Mean Variance	-	The mean or average of the absolute values of each data point's variance.
Probability of Success	-	The chances of being successful out of a number of tries.

$$P = \frac{s}{s+f}$$

IMAGINARY AND COMPLEX NUMBERS

This chapter will cover the definitions and rules for the application of imaginary and complex numbers.

EO 2.1 **STATE the definition of an imaginary number.**

EO 2.2 **STATE the definition of a complex number.**

EO 2.3 **APPLY the arithmetic operations of addition, subtraction, and multiplication, and division to complex numbers.**

Imaginary and complex numbers are entirely different from any kind of number used up to this point. These numbers are generated when solving some quadratic and higher degree equations. Imaginary and complex numbers become important in the study of electricity; especially in the study of alternating current circuits.

Imaginary Numbers

Imaginary numbers result when a mathematical operation yields the square root of a negative number. For example, in solving the quadratic equation $x^2 + 25 = 0$, the solution yields $x^2 = -25$. Thus, the roots of the equation are $x = \pm\sqrt{-25}$. The square root of (-25) is called an imaginary number. Actually, any even root (i.e. square root, 4th root, 6th root, etc.) of a negative number is called an imaginary number. All other numbers are called real numbers. The name "imaginary" may be somewhat misleading since imaginary numbers actually exist and can be used in mathematical operations. They can be added, subtracted, multiplied, and divided.

Imaginary numbers are written in a form different from real numbers. Since they are radicals, they can be simplified by factoring. Thus, the imaginary number $\sqrt{-25}$ equals $\sqrt{(25)(-1)}$, which equals $5\sqrt{-1}$. Similarly, $\sqrt{-9}$ equals $\sqrt{(9)(-1)}$, which equals $3\sqrt{-1}$. All imaginary numbers can be simplified in this way. They can be written as the product of a real number and $\sqrt{-1}$. In order to further simplify writing imaginary numbers, the imaginary unit i is defined as $\sqrt{-1}$. Thus, the imaginary number, $\sqrt{-25}$, which equals $5\sqrt{-1}$, is written as $5i$, and the imaginary number, $\sqrt{-9}$, which equals $3\sqrt{-1}$, is written $3i$. In using imaginary numbers in electricity, the imaginary unit is often represented by j, instead of i, since i is the common notation for electrical current.

Imaginary numbers are added or subtracted by writing them using the imaginary unit *i* and then adding or subtracting the real number coefficients of *i*. They are added or subtracted like algebraic terms in which the imaginary unit *i* is treated like a literal number. Thus, $\sqrt{-25}$ and $\sqrt{-9}$ are added by writing them as $5i$ and $3i$ and adding them like algebraic terms. The result is $8i$ which equals $8\sqrt{-1}$ or $\sqrt{-64}$. Similarly, $\sqrt{-9}$ subtracted from $\sqrt{-25}$ equals $3i$ subtracted from $5i$ which equals $2i$ or $2\sqrt{-1}$ or $\sqrt{-4}$.

Example:

Combine the following imaginary numbers:

Solution:

$$\sqrt{-16} + \sqrt{-36} - \sqrt{-49} - \sqrt{-1} =$$

$$\begin{aligned}\sqrt{-16} + \sqrt{-36} - \sqrt{-49} - \sqrt{-1} &= 4i + 6i - 7i - i \\ &= 10i - 8i \\ &= 2i\end{aligned}$$

Thus, the result is $2i = 2\sqrt{-1} = \sqrt{-4}$

Imaginary numbers are multiplied or divided by writing them using the imaginary unit *i*, and then multiplying or dividing them like algebraic terms. However, there are several basic relationships which must also be used to multiply or divide imaginary numbers.

$$i^2 = (i)(i) = (\sqrt{-1})(\sqrt{-1}) = -1$$
$$i^3 = (i^2)(i) = (-1)(i) = -i$$
$$i^4 = (i^2)(i^2) = (-1)(-1) = +1$$

Using these basic relationships, for example, $(\sqrt{-25})(\sqrt{-4})$ equals $(5i)(2i)$ which equals $10i^2$. But, i^2 equals -1. Thus, $10i^2$ equals (10)(-1) which equals -10.

Any square root has two roots, i.e., a statement $x^2 = 25$ is a quadratic and has roots

$x = \pm 5$ since $+5^2 = 25$ and $(-5) \times (-5) = 25$.

Similarly,

$\sqrt{-25} = \pm 5i$
$\sqrt{-4} = \pm 2i$
and
$\sqrt{-25}\ \sqrt{-4} = \pm 10.$

Example 1:

Multiply $\sqrt{-2}$ and $\sqrt{-32}$.

Solution:

$$(\sqrt{-2}\,)(\sqrt{-32}\,) = (\sqrt{2}\,i)(\sqrt{32}\,i)$$

$$= \sqrt{(2)(32)}\,i^2$$

$$= \sqrt{64}\,(-1)$$

$$= 8\,(-1)$$

$$= -8$$

Example 2:

Divide $\sqrt{-48}$ by $\sqrt{-3}$.

Solution:

$$\frac{\sqrt{-48}}{\sqrt{-3}} = \frac{\sqrt{48}i}{\sqrt{3}i}$$

$$= \sqrt{\frac{48}{3}}$$

$$= \sqrt{16}$$

$$= 4$$

Complex Numbers

Complex numbers are numbers which consist of a real part and an imaginary part. The solution of some quadratic and higher degree equations results in complex numbers. For example, the roots of the quadratic equation, $x^2 - 4x + 13 = 0$, are complex numbers. Using the quadratic formula yields two complex numbers as roots.

$$x = \frac{-b \pm \sqrt{b^2 - 4ac}}{2a}$$

$$x = \frac{4 \pm \sqrt{16 - 52}}{2}$$

$$x = \frac{4 \pm \sqrt{-36}}{2}$$

$$x = \frac{4 \pm 6i}{2}$$

$$x = 2 \pm 3i$$

The two roots are $2 + 3i$ and $2 - 3i$; they are both complex numbers. 2 is the real part; $+3i$ and $-3i$ are the imaginary parts. The general form of a complex number is $a + bi$, in which "a" represents the real part and "bi" represents the imaginary part.

Complex numbers are added, subtracted, multiplied, and divided like algebraic binomials. Thus, the sum of the two complex numbers, $7 + 5i$ and $2 + 3i$ is $9 + 8i$, and $7 + 5i$ minus $2 + 3i$, is $5 + 2i$. Similarly, the product of $7 + 5i$ and $2 + 3i$ is $14 + 31i + 15i^2$. But i^2 equals -1. Thus, the product is $14 + 31i + 15(-1)$ which equals $-1 + 31i$.

Example 1:

Combine the following complex numbers:

$$(4 + 3i) + (8 - 2i) - (7 + 3i) =$$

Solution:

$$\begin{aligned}(4 + 3i) + (8 - 2i) - (7 + 3i) &= (4 + 8 - 7) + (3 - 2 - 3)i \\ &= 5 - 2i\end{aligned}$$

Example 2:

Multiply the following complex numbers:

$(3 + 5i)(6 - 2i) =$

Solution:

$$
\begin{aligned}
(3 + 5i)(6 - 2i) &= 18 + 30i - 6i - 10i^2 \\
&= 18 + 24i - 10(-1) \\
&= 28 + 24i
\end{aligned}
$$

Example 3:

Divide $(6+8i)$ by 2.

Solution:

$$
\begin{aligned}
\frac{6 + 8i}{2} &= \frac{6}{2} + \frac{8}{2}i \\
&= 3 + 4i
\end{aligned}
$$

A difficulty occurs when dividing one complex number by another complex number. To get around this difficulty, one must eliminate the imaginary portion of the complex number from the denominator, when the division is written as a fraction. This is accomplished by multiplying the numerator and denominator by the *conjugate* form of the denominator. The *conjugate* of a complex number is that complex number written with the opposite sign for the imaginary part. For example, the conjugate of $4+5i$ is $4-5i$.

This method is best demonstrated by example.

Example: $(4 + 8i) \div (2 - 4i)$

Solution:

$$
\begin{aligned}
\frac{4 + 8i}{2 - 4i} \cdot \frac{2 + 4i}{2 + 4i} &= \frac{8 + 32i + 32i^2}{4 - 16i^2} \\
&= \frac{8 + 32i + 32(-1)}{4 - 16(-1)} \\
&= \frac{-24 + 32i}{20} \\
&= -\frac{6}{5} + \frac{8}{5}i
\end{aligned}
$$

Summary

The important information from this chapter is summartized below.

Imaginary and Complex Numbers Summary

Imaginary Number

- An Imaginary number is the square root of a negative number.

Complex Number

- A complex number is any number that contains both a real and imaginary term.

Addition and Subtraction of Complex Numbers

- Add/subtract the real terms together, and add/subtract the imaginary terms of each complex number together. The result will be a complex number.

Multiplication of Complex Numbers

- Treat each complex number as an algebraic term and multiply/divide using rules of algebra. The result will be a complex number.

Division of Complex Numbers

- Put division in fraction form and multiply numerator and denominator by the conjugate of the denominator.

Rules of the Imaginary Number i

- $i^2 = (i)(i) = -1$
  $i^3 = (i^2)(i) = (-1)(i) = -i$
  $i^4 = (i^2)(i^2) = (-1)(-1) = +1$

MATRICES AND DETERMINANTS

This chapter will explain the idea of matrices and determinate and the rules needed to apply matrices in the solution of simultaneous equations.

EO 3.1 DETERMINE the dimensions of a given matrix.

EO 3.2 SOLVE a given set of equations using Cramer's Rule.

In the real world, many times the solution to a problem containing a large number of variables is required. In both physics and electrical circuit theory, many problems will be encountered which contain multiple simultaneous equations with multiple unknowns. These equations can be solved using the standard approach of eliminating the variables or by one of the other methods. This can be difficult and time-consuming. To avoid this problem, and easily solve families of equations containing multiple unknowns, a type of math was developed called Matrix theory.

Once the terminology and basic manipulations of matrices are understood, matrices can be used to readily solve large complex systems of equations.

The Matrix

We define a matrix as any rectangular array of numbers. Examples of matrices may be formed from the coefficients and constants of a system of linear equations: that is,

$$2x - 4y = 7$$
$$3x + y = 16$$

can be written as follows.

$$\begin{bmatrix} 2 & -4 & 7 \\ 3 & 1 & 16 \end{bmatrix}$$

The numbers used in the matrix are called elements. In the example given, we have three columns and two rows of elements. The number of rows and columns are used to determine the dimensions of the matrix. In our example, the dimensions of the matrix are 2 x 3, having 2 rows and 3 columns of elements. In general, the dimensions of a matrix which have *m* rows and *n* columns is called an *m* x *n* matrix.

A matrix with only a single row or a single column is called either a row or a column matrix. A matrix which has the same number of rows as columns is called a square matrix. Examples of matrices and their dimensions are as follows:

$$\begin{bmatrix} 1 & 7 & 6 \\ 2 & 4 & 8 \end{bmatrix} = 2 \text{ x } 3$$

$$\begin{bmatrix} 1 & 7 \\ 6 & 2 \\ 3 & 5 \end{bmatrix} = 3 \text{ x } 2$$

$$\begin{bmatrix} 3 \\ 2 \\ 1 \end{bmatrix} = 3 \text{ x } 1$$

We will use capital letters to describe matrices. We will also include subscripts to give the dimensions.

$$A_{2\times3} = \begin{bmatrix} 1 & 3 & 3 \\ 5 & 6 & 7 \end{bmatrix}$$

Two matrices are said to be equal if, and only if, they have the same dimensions, and their corresponding elements are equal. The following are all equal matrices:

$$\begin{bmatrix} 0 & 1 \\ 2 & 4 \end{bmatrix} = \begin{bmatrix} 0 & 1 \\ 2 & 4 \end{bmatrix} = \begin{bmatrix} 0 & \frac{1}{1} \\ \frac{6}{3} & 4 \end{bmatrix}$$

Addition of Matrices

Matrices may only be added if they both have the same dimensions. To add two matrices, each element is added to its corresponding element. The sum matrix has the same dimensions as the two being added.

Example:

Add matrix A to matrix B.

$$A = \begin{bmatrix} 6 & 2 & 6 \\ -1 & 3 & 0 \end{bmatrix} \qquad B = \begin{bmatrix} 2 & 1 & 3 \\ 0 & -3 & 6 \end{bmatrix}$$

Solution:

$$A + B = \begin{bmatrix} 6+2 & 2+1 & 6+3 \\ -1+0 & 3-3 & 0+6 \end{bmatrix}$$

$$= \begin{bmatrix} 8 & 3 & 9 \\ -1 & 0 & 6 \end{bmatrix}$$

Multiplication of a Scalar and a Matrix

When multiplying a matrix by a scalar (or number), we write "scalar K times matrix A." Each element of the matrix is multiplied by the scalar. By example:

$$K = 3 \quad \text{and} \quad A = \begin{bmatrix} 2 & 3 \\ 1 & 7 \end{bmatrix}$$

then

$$3 \text{ x } A = 3\begin{bmatrix} 2 & 3 \\ 1 & 7 \end{bmatrix}$$

$$= \begin{bmatrix} 2 \cdot 3 & 3 \cdot 3 \\ 1 \cdot 3 & 7 \cdot 3 \end{bmatrix}$$

$$= \begin{bmatrix} 6 & 9 \\ 3 & 21 \end{bmatrix}$$

Multiplication of a Matrix by a Matrix

To multiply two matrices, the first matrix must have the same number of rows (*m*) as the second matrix has columns (*n*). In other words, *m* of the first matrix must equal *n* of the second matrix. For example, a 2 x 1 matrix can be multiplied by a 1 x 2 matrix,

$$\begin{bmatrix} x \\ y \end{bmatrix} \begin{bmatrix} a & b \end{bmatrix} = \begin{bmatrix} ax & bx \\ ay & by \end{bmatrix}$$

or a 2 x 2 matrix can be multiplied by a 2 x 2. If an *m* x *n* matrix is multiplied by an *n* x *p* matrix, then the resulting matrix is an *m* x *p* matrix. For example, if a 2 x 1 and a 1 x 2 are multiplied, the result will be a 2 x 2. If a 2 x 2 and a 2 x 2 are multiplied, the result will be a 2 x 2.

To multiply two matrices, the following pattern is used:

$$A = \begin{bmatrix} a & b \\ c & d \end{bmatrix} \qquad B = \begin{bmatrix} w & x \\ y & z \end{bmatrix}$$

$$C = A \cdot B = \begin{bmatrix} aw+by & ax+bz \\ cw+dy & cx+dz \end{bmatrix}$$

In general terms, a matrix C which is a product of two matrices, A and B, will have elements given by the following.

$$c_{ij} = a_{i1}b_{1j} + a_{j2}b_{2j} + + + \ldots + a_{in}b_{nj}$$

where

i = ith row
j = jth column

Example:

Multiply the matrices *A* x *B*.

$$A = \begin{bmatrix} 1 & 2 \\ 3 & 4 \end{bmatrix} \qquad B = \begin{bmatrix} 3 & 5 \\ 0 & 6 \end{bmatrix}$$

Solution:

$$A \cdot B = \begin{bmatrix} (1x3)+(2x0) & (1x5)+(2x6) \\ (3x3)+(4x0) & (3x5)+(4x6) \end{bmatrix}$$

$$= \begin{bmatrix} 3+0 & 5+12 \\ 9+0 & 15+24 \end{bmatrix}$$

$$= \begin{bmatrix} 3 & 17 \\ 9 & 39 \end{bmatrix}$$

It should be noted that the multiplication of matrices is not usually commutative.

The Determinant

Square matrixes have a property called a determinant. When a determinant of a matrix is written, it is symbolized by vertical bars rather than brackets around the numbers. This differentiates the determinant from a matrix. The determinant of a matrix is the reduction of the matrix to a single scalar number. The determinant of a matrix is found by "expanding" the matrix. There are several methods of "expanding" a matrix and calculating it's determinant. In this lesson, we will only look at a method called "expansion by minors."

Before a large matrix determinant can be calculated, we must learn how to calculate the determinant of a 2 x 2 matrix. By definition, the determinant of a 2 x 2 matrix is calculated as follows:

$$A = \begin{vmatrix} a & b \\ c & d \end{vmatrix} = ad - bc$$

Example: Find the determinant of A.

$$A = \begin{vmatrix} 6 & 2 \\ -1 & 3 \end{vmatrix}$$

Solution:

$$\begin{aligned} A &= (6 \cdot 3) - (-1 \cdot 2) \\ &= 18 - (-2) \\ &= 18 + 2 \\ &= 20 \end{aligned}$$

To expand a matrix larger than a 2 x 2 requires that it be simplified down to several 2 x 2 matrices, which can then be solved for their determinant. It is easiest to explain the process by example.

Given the 3 x 3 matrix:

$$\begin{matrix} 1 & 3 & 1 \\ 4 & 1 & 2 \\ 5 & 6 & 3 \end{matrix}$$

Any single row or column is picked. In this example, column one is selected. The matrix will be expanded using the elements from the first column. Each of the elements in the selected column will be multiplied by its minor starting with the first element in the column (1). A line is then drawn through all the elements in the same row and column as 1. Since this is a 3 x 3 matrix, that leaves a minor or 2 x 2 determinant. This resulting 2 x 2 determinant is called the minor of the element in the first row first column.

$$\begin{bmatrix} 1 & 3 & 1 \\ 4 & 1 & 2 \\ 5 & 6 & 3 \end{bmatrix}$$

$$1 \begin{vmatrix} 1 & 2 \\ 6 & 3 \end{vmatrix}$$

The minor of element 4 is:

$$\begin{bmatrix} 1 & 3 & 1 \\ 4 & 1 & 2 \\ 5 & 6 & 3 \end{bmatrix}$$

$$4 \begin{vmatrix} 3 & 1 \\ 6 & 3 \end{vmatrix}$$

The minor of element 5 is:

$$\begin{bmatrix} 1 & 3 & 1 \\ 4 & 1 & 2 \\ 5 & 6 & 3 \end{bmatrix}$$

$$5 \begin{vmatrix} 3 & 1 \\ 1 & 2 \end{vmatrix}$$

Each element is given a sign based on its position in the original determinant.

$$\begin{vmatrix} + & - & + \\ - & + & - \\ + & - & + \end{vmatrix}$$

The sign is positive (negative) if the sum of the row plus the column for the element is even (odd). This pattern can be expanded or reduced to any size determinant. The positive and negative signs are just alternated.

Each minor is now multiplied by its signed element and the determinant of the resulting 2 x 2 calculated.

$$1\begin{bmatrix} 1 & 2 \\ 6 & 3 \end{bmatrix} = 1\left[(1 \cdot 3) - (2 \cdot 6)\right] = 3 - (12) = -9$$

$$-4\begin{bmatrix} 3 & 1 \\ 6 & 3 \end{bmatrix} = -4\left[(3 \cdot 3) - (1 \cdot 6)\right] = -4\left[9 - 6\right] = -12$$

$$5\begin{bmatrix} 3 & 1 \\ 1 & 2 \end{bmatrix} = 5\left[(3 \cdot 2) - (1 \cdot 1)\right] = 5\left[6 - 1\right] = 25$$

Determinant = (-9) + (-12) + 25 = 4

Example:

Find the determinant of the following 3 x 3 matrix, expanding about row 1.

$$\begin{vmatrix} 3 & 1 & 2 \\ 4 & 5 & 6 \\ 0 & 1 & 4 \end{vmatrix}$$

Solution:

First Minor

$$\begin{bmatrix} 3 & 1 & 3 \\ 4 & 5 & 6 \\ 0 & 1 & 4 \end{bmatrix} = 3\begin{vmatrix} 5 & 6 \\ 1 & 4 \end{vmatrix} = 3\,(20 - 6) = 3\,(14) = 42$$

Second Minor

$$\begin{bmatrix} 3 & 1 & 3 \\ 4 & 5 & 6 \\ 0 & 1 & 4 \end{bmatrix} = -1\begin{vmatrix} 4 & 6 \\ 0 & 4 \end{vmatrix} = -1\,(16 - 0) = -1\,(16) = -16$$

Third Minor

$$\begin{bmatrix} 3 & 1 & 3 \\ 4 & 5 & 6 \\ 0 & 1 & 4 \end{bmatrix} = 3\begin{vmatrix} 4 & 5 \\ 0 & 1 \end{vmatrix} = 3\,(4 - 0) = 3\,(4) = 12$$

$$\text{Determinant} = 42 + (-16) + (12) = 38$$

Using Matrices to Solve System of Linear Equation (Cramer's Rule)

Matrices and their determinant can be used to solve a system of equations. This method becomes especially attractive when large numbers of unknowns are involved. But the method is still useful in solving algebraic equations containing two and three unknowns.

In part one of this chapter, it was shown that equations could be organized such that their coefficients could be written as a matrix.

$$ax + by = c$$
$$ex + fy = g$$

where:

x and y are variables
a, b, e, and f are the coefficients
c and g are constants

The equations can be rewritten in matrix form as follows:

$$\begin{bmatrix} a & b \\ e & f \end{bmatrix} \begin{bmatrix} x \\ y \end{bmatrix} = \begin{bmatrix} c \\ g \end{bmatrix}$$

To solve for each variable, the matrix containing the constants (c,g) is substituted in place of the column containing the coefficients of the variable that we want to solve for (a,e or b,f). This new matrix is divided by the original coefficient matrix. This process is call "Cramer's Rule."

Example:

In the above problem to solve for x,

$$x = \frac{\begin{vmatrix} c & b \\ g & f \end{vmatrix}}{\begin{vmatrix} c & b \\ g & f \end{vmatrix}}$$

and to solve for y,

$$y = \frac{\begin{vmatrix} a & c \\ e & g \end{vmatrix}}{\begin{vmatrix} a & b \\ e & f \end{vmatrix}}$$

Example:

Solve the following two equations:

$$x + 2y = 4$$
$$-x + 3y = 1$$

Solution:

$$x = \frac{\begin{vmatrix} 4 & 2 \\ 1 & 3 \end{vmatrix}}{\begin{vmatrix} 1 & 2 \\ -1 & 3 \end{vmatrix}}$$

$$y = \frac{\begin{vmatrix} 1 & 4 \\ -1 & 1 \end{vmatrix}}{\begin{vmatrix} 1 & 2 \\ -1 & 3 \end{vmatrix}}$$

solving each 2 x 2 for its determinant,

$$x = \frac{[(4 \cdot 3) - (1 \cdot 2)]}{[(1 \cdot 3) - (-1 \cdot 2)]} = \frac{12 - 2}{3 + 2} = \frac{10}{5} = 2$$

$$y = \frac{[(1 \cdot 1) - (-1 \cdot 4)]}{[(1 \cdot 3) - (-1 \cdot 2)]} = \frac{1 + 4}{3 + 2} = \frac{5}{5} = 1$$

$$x = 2 \quad \text{and} \quad y = 1$$

A 3 x 3 is solved by using the same logic, except each 3 x 3 must be expanded by minors to solve for the determinant.

Example:

Given the following three equations, solve for the three unknowns.

$2x + 3y - z = 2$
$x - 2y + 2z = -10$
$3x + y - 2z = 1$

Solution:

$$x = \frac{\begin{vmatrix} 2 & 3 & -1 \\ -10 & -2 & 2 \\ 1 & 1 & -2 \end{vmatrix}}{\begin{vmatrix} 2 & 3 & -1 \\ 1 & -2 & 2 \\ 3 & 1 & -2 \end{vmatrix}}$$

$$y = \frac{\begin{vmatrix} 2 & 2 & -1 \\ 1 & -10 & 2 \\ 3 & 1 & -2 \end{vmatrix}}{\begin{vmatrix} 2 & 3 & -1 \\ 1 & -2 & 2 \\ 3 & 1 & -2 \end{vmatrix}}$$

$$z = \frac{\begin{vmatrix} 2 & 3 & 2 \\ 1 & -2 & -10 \\ 3 & 1 & 1 \end{vmatrix}}{\begin{vmatrix} 2 & 3 & -1 \\ 1 & -2 & 2 \\ 3 & 1 & -2 \end{vmatrix}}$$

Expanding the top matrix for x using the elements in the bottom row gives:

$$1\begin{bmatrix} 3 & -1 \\ -2 & 2 \end{bmatrix} + (-1)\begin{bmatrix} 2 & -1 \\ -10 & 2 \end{bmatrix} + (-2)\begin{bmatrix} 2 & 3 \\ -10 & -2 \end{bmatrix} =$$

$$1(6-2) + (-1)(4-10) + (-2)(-4+30) =$$

$$4 + 6 - 52 = -42$$

Expanding the bottom matrix for x using the elements in the first column gives:

$$2\begin{bmatrix} -2 & 2 \\ 1 & -2 \end{bmatrix} + (-1)\begin{bmatrix} 3 & -1 \\ 1 & -2 \end{bmatrix} + 3\begin{bmatrix} 3 & -1 \\ -2 & 2 \end{bmatrix} =$$

$$2(4 - 2) + (-1)(-6 + 1) + 3(6 - 2) =$$

$$4 + 5 + 12 = 21$$

This gives:

$$x = \frac{-42}{21} = -2$$

y and z can be expanded using the same method.

$$y = 1$$
$$z = -3$$

Summary

The use of matrices and determinants is summarized below.

Matrices and Determinant Summary

The dimensions of a matrix are given as m x n, where m = number of rows and n = number of columns.

The use of determinants and matrices to solve linear equations is done by:

- placing the coefficients and constants into a determinant format.
- substituting the constants in place of the coefficients of the variable to be solved for.
- dividing the new-formed substituted determinant by the original determinant of coefficients.
- expanding the determinant.

CALCULUS

Many practical problems can be solved using arithmetic and algebra; however, many other practical problems involve quantities that cannot be adequately described using numbers which have fixed values.

EO 4.1 STATE the graphical definition of a derivative.

EO 4.2 STATE the graphical definition of an integral.

Dynamic Systems

Arithmetic involves numbers that have fixed values. Algebra involves both literal and arithmetic numbers. Although the literal numbers in algebraic problems can change value from one calculation to the next, they also have fixed values in a given calculation. When a weight is dropped and allowed to fall freely, its velocity changes continually. The electric current in an alternating current circuit changes continually. Both of these quantities have a different value at successive instants of time. Physical systems that involve quantities that change continually are called dynamic systems. The solution of problems involving dynamic systems often involves mathematical techniques different from those described in arithmetic and algebra. Calculus involves all the same mathematical techniques involved in arithmetic and algebra, such as addition, subtraction, multiplication, division, equations, and functions, but it also involves several other techniques. These techniques are not difficult to understand because they can be developed using familiar physical systems, but they do involve new ideas and terminology.

There are many dynamic systems encountered in nuclear facility work. The decay of radioactive materials, the startup of a reactor, and a power change on a turbine generator all involve quantities which change continually. An analysis of these dynamic systems involves calculus. Although the operation of a nuclear facility does not require a detailed understanding of calculus, it is most helpful to understand certain of the basic ideas and terminology involved. These ideas and terminology are encountered frequently, and a brief introduction to the basic ideas and terminology of the mathematics of dynamic systems is discussed in this chapter.

Differentials and Derivatives

One of the most commonly encountered applications of the mathematics of dynamic systems involves the relationship between position and time for a moving object. Figure 2 represents an object moving in a straight line from position P_1 to position P_2. The distance to P_1 from a fixed reference point, point 0, along the line of travel is represented by S_1; the distance to P_2 from point 0 by S_2.

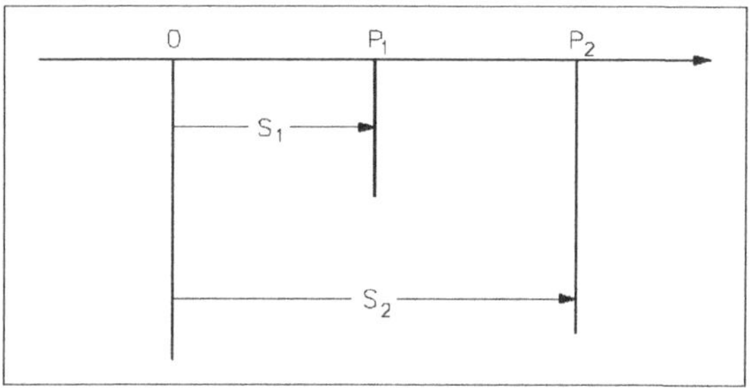

Figure 2 Motion Between Two Points

If the time recorded by a clock, when the object is at position P_1 is t_1, and if the time when the object is at position P_2 is t_2, then the average velocity of the object between points P_1 and P_2 equals the distance traveled, divided by the elapsed time.

$$V_{av} = \frac{S_2 - S_1}{t_2 - t_1} \tag{5-1}$$

If positions P_1 and P_2 are close together, the distance traveled and the elapsed time are small. The symbol Δ, the Greek letter delta, is used to denote changes in quantities. Thus, the average velocity when positions P_1 and P_2 are close together is often written using deltas.

$$V_{av} = \frac{\Delta S}{\Delta t} = \frac{S_2 - S_1}{t_2 - t_1} \tag{5-2}$$

Although the average velocity is often an important quantity, in many cases it is necessary to know the velocity at a given instant of time. This velocity, called the instantaneous velocity, is not the same as the average velocity, unless the velocity is not changing with time.

Using the graph of displacement, S, versus time, t, in Figure 3, we will try to describe the concept of the derivative.

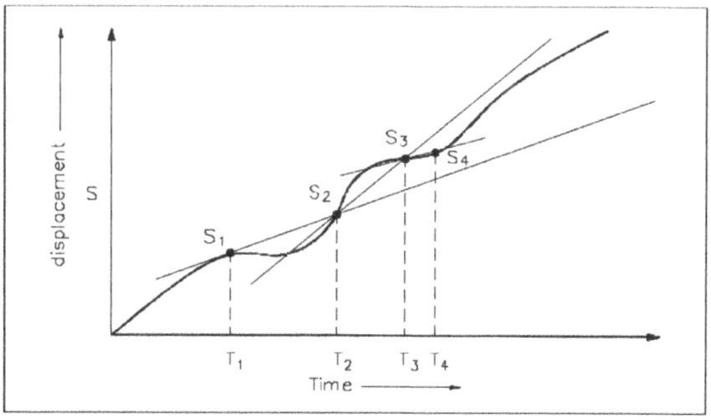

Figure 3 Displacement Versus Time

Using equation 5-1 we find the average velocity from S_1 to S_2 is $\frac{S_2 - S_1}{t_2 - t_1}$. If we connect the points S_1 and S_2 by a straight line we see it does not accurately reflect the slope of the curved line through all the points between S_1 and S_2. Similarly, if we look at the average velocity between time t_2 and t_3 (a smaller period of time), we see the straight line connecting S_2 and S_3 more closely follows the curved line. Assuming the time between t_3 and t_4 is less than between t_2 and t_3, the straight line connecting S_3 and S_4 very closely approximates the curved line between S_3 and S_4.

As we further decrease the time interval between successive points, the expression $\frac{\Delta S}{\Delta t}$ more closely approximates the slope of the displacement curve. As $\Delta t \to 0$, $\frac{\Delta S}{\Delta t}$ approaches the instantaneous velocity. The expression for the derivative (in this case the slope of the displacement curve) can be written $\frac{dS}{dt} = \lim_{\Delta t \to 0} \frac{\Delta S}{\Delta t}$. In words, this expression would be "the derivative of S with respect to time (t) is the limit of $\frac{\Delta S}{\Delta t}$ as Δt approaches 0."

$$V = \frac{ds}{dt} = \lim_{\Delta t \to 0} \frac{\Delta s}{\Delta t} \qquad (5\text{-}3)$$

The symbols *ds* and *dt* are not products of *d* and *s*, or of *d* and *t*, as in algebra. Each represents a single quantity. They are pronounced "dee-ess" and "dee-tee," respectively. These expressions and the quantities they represent are called differentials. Thus, *ds* is the differential of *s* and *dt* is the differential of *t*. These expressions represent incremental changes, where *ds* represents an incremental change in distance *s*, and *dt* represents an incremental change in time *t*.

The combined expression *ds/dt* is called a derivative; it is the derivative of *s* with respect to *t*. It is read as "dee-ess dee-tee." *dz/dx* is the derivative of *z* with respect to *x*; it is read as "dee-zee dee-ex." In simplest terms, a derivative expresses the rate of change of one quantity with respect to another. Thus, *ds/dt* is the rate of change of distance with respect to time. Referring to figure 3, the derivative *ds/dt* is the instantaneous velocity at any chosen point along the curve. This value of instantaneous velocity is numerically equal to the slope of the curve at that chosen point.

While the equation for instantaneous velocity, $V = ds/dt$, may seem like a complicated expression, it is a familiar relationship. Instantaneous velocity is precisely the value given by the speedometer of a moving car. Thus, the speedometer gives the value of the rate of change of distance with respect to time; it gives the derivative of *s* with respect to *t*; i.e. it gives the value of *ds/dt*.

The ideas of differentials and derivatives are fundamental to the application of mathematics to dynamic systems. They are used not only to express relationships among distance traveled, elapsed time and velocity, but also to express relationships among many different physical quantities. One of the most important parts of understanding these ideas is having a physical interpretation of their meaning. For example, when a relationship is written using a differential or a derivative, the physical meaning in terms of incremental changes or rates of change should be readily understood.

When expressions are written using deltas, they can be understood in terms of changes. Thus, the expression ΔT, where T is the symbol for temperature, represents a change in temperature. As previously discussed, a lower case delta, d, is used to represent very small changes. Thus, dT represents a very small change in temperature. The fractional change in a physical quantity is the change divided by the value of the quantity. Thus, dT is an incremental change in temperature, and dT/T is a fractional change in temperature. When expressions are written as derivatives, they can be understood in terms of rates of change. Thus, dT/dt is the rate of change of temperature with respect to time.

Examples:

1. Interpret the expression $\Delta V/V$, and write it in terms of a differential. $\Delta V/V$ expresses the fractional change of velocity. It is the change in velocity divided by the velocity. It can be written as a differential when ΔV is taken as an incremental change.

$$\frac{\Delta V}{V} \text{ may be written as } \frac{dV}{V}$$

2. Give the physical interpretation of the following equation relating the work W done when a force F moves a body through a distance x.

$$dW = Fdx$$

This equation includes the differentials dW and dx which can be interpreted in terms of incremental changes. The incremental amount of work done equals the force applied multiplied by the incremental distance moved.

3. Give the physical interpretation of the following equation relating the force, F, applied to an object, its mass m, its instantaneous velocity v and time t.

$$F = m\frac{dv}{dt}$$

This equation includes the derivative dv/dt; the derivative of the velocity with respect to time. It is the rate of change of velocity with respect to time. The force applied to an object equals the mass of the object multiplied by the rate of change of velocity with respect to time.

4. Give the physical interpretation of the following equation relating the acceleration a, the velocity v, and the time t.

$$a = \frac{dv}{dt}$$

This equation includes the derivative dv/dt; the derivative of the velocity with respect to time. It is a rate of change. The acceleration equals the rate of change of velocity with respect to time.

Graphical Understanding of Derivatives

A function expresses a relationship between two or more variables. For example, the distance traveled by a moving body is a function of the body's velocity and the elapsed time. When a functional relationship is presented in graphical form, an important understanding of the meaning of derivatives can be developed.

Figure 4 is a graph of the distance traveled by an object as a function of the elapsed time. The functional relationship shown is given by the following equation:

$$s = 40t \qquad (5\text{-}4)$$

The instantaneous velocity v, which is the velocity at a given instant of time, equals the derivative of the distance traveled with respect to time, ds/dt. It is the rate of change of s with respect to t.

The value of the derivative *ds/dt* for the case plotted in Figure 4 can be understood by considering small changes in the two variables *s* and *t*.

$$\frac{\Delta s}{\Delta t} = \frac{(s + \Delta s) - s}{(t + \Delta t) - t}$$

The values of $(s + \Delta s)$ and s in terms of $(t + \Delta t)$ and t, using Equation 5-4 can now be substituted into this expression. At time t, $s = 40t$; at time $t + \Delta t$, $s + \Delta s = 40(t + \Delta t)$.

$$\frac{\Delta s}{\Delta t} = \frac{40(t + \Delta t) - 40t}{(t + \Delta t) - t}$$

$$\frac{\Delta s}{\Delta t} = \frac{40t + 40(\Delta t) - 40t}{t + \Delta t - t}$$

$$\frac{\Delta s}{\Delta t} = \frac{40(\Delta t)}{\Delta t}$$

$$\frac{\Delta s}{\Delta t} = 40$$

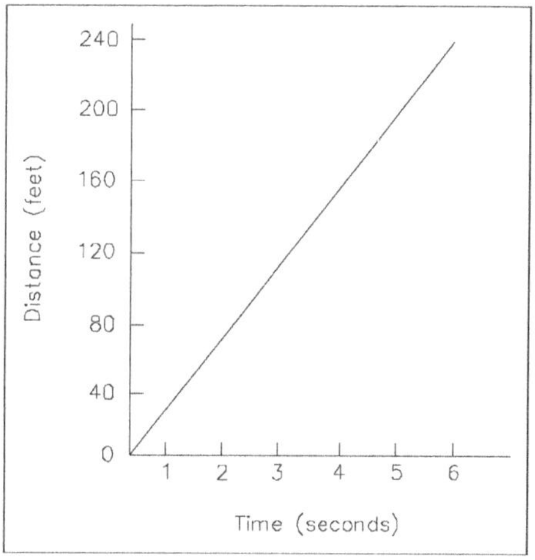

Figure 4 Graph of Distance vs. Time

The value of the derivative *ds/dt* in the case plotted in Figure 4 is a constant. It equals 40 ft/s. In the discussion of graphing, the slope of a straight line on a graph was defined as the change in y, Δy, divided by the change in x, Δx. The slope of the line in Figure 4 is $\Delta s/\Delta t$ which, in this case, is the value of the derivative *ds/dt*. Thus, derivatives of functions can be interpreted in terms of the slope of the graphical plot of the function. Since the velocity equals the derivative of the distance s with respect to time t, *ds/dt*, and since this derivative equals the slope of the plot of distance versus time, the velocity can be visualized as the slope of the graphical plot of distance versus time.

For the case shown in Figure 4, the velocity is constant. Figure 5 is another graph of the distance traveled by an object as a function of the elapsed time. In this case the velocity is not constant. The functional relationship shown is given by the following equation:

$$s = 10t^2 \qquad (5\text{-}5)$$

The instantaneous velocity again equals the value of the derivative *ds/dt*. This value is changing with time. However, the instantaneous velocity at any specified time can be determined. First, small changes in *s* and *t* are considered.

$$\frac{\Delta s}{\Delta t} = \frac{(s + \Delta s) - s}{(t + \Delta t) - t}$$

The values of $(s + \Delta s)$ and s in terms of $(t + \Delta t)$ and t using Equation 5-5, can then be substituted into this expression. At time t, $s = 10t^2$; at time $t + \Delta t$, $s + \Delta s = 10(t + \Delta t)^2$. The value of $(t + \Delta t)^2$ equals $t^2 + 2t(\Delta t) + (\Delta t)^2$; however, for incremental values of Δt, the term $(\Delta t)^2$ is so small, it can be neglected. Thus, $(t + \Delta t)^2 = t^2 + 2t(\Delta t)$.

$$\frac{\Delta s}{\Delta t} = \frac{10[t^2 + 2t(\Delta t)] - 10t^2}{(t + \Delta t) - t}$$

$$\frac{\Delta s}{\Delta t} = \frac{10t^2 + 20t(\Delta t)] - 10t^2}{t + \Delta t - t}$$

$$\frac{\Delta s}{\Delta t} = 20t$$

The value of the derivative *ds/dt* in the case plotted in Figure 5 equals $20t$. Thus, at time $t = 1$ *s*, the instantaneous velocity equals 20 ft/s; at time $t = 2$ *s*, the velocity equals 40 ft/s, and so on.

When the graph of a function is not a straight line, the slope of the plot is different at different points. The slope of a curve at any point is defined as the slope of a line drawn tangent to the curve at that point. Figure 6 shows a line drawn tangent to a curve. A tangent line is a line that touches the curve at only one point. The line *AB* is tangent to the

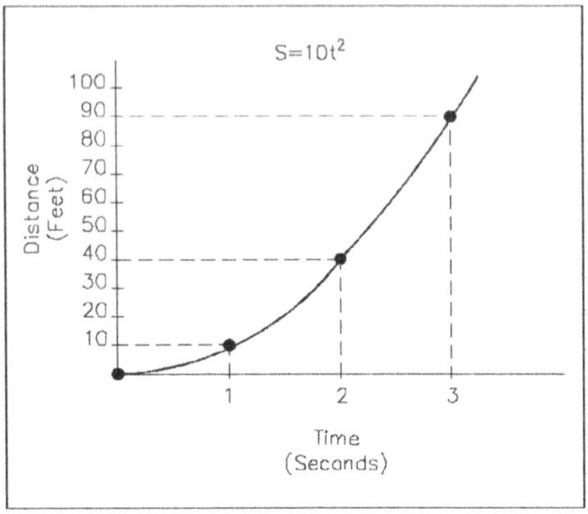

Figure 5 Graph of Distance vs. Time

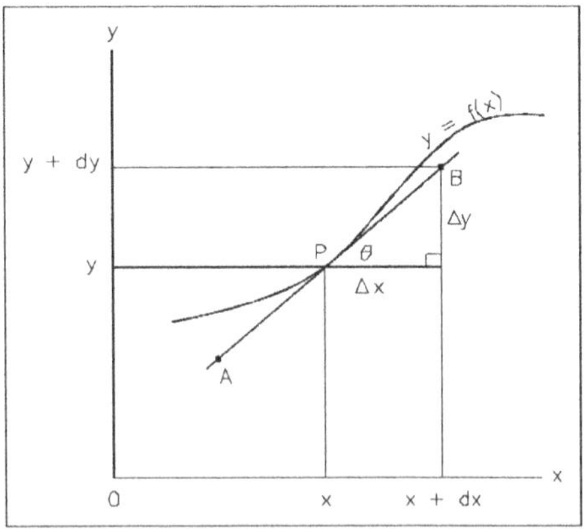

Figure 6 Slope of a Curve

curve $y = f(x)$ at point P.

The tangent line has the slope of the curve dy/dx, where, θ is the angle between the tangent line AB and a line parallel to the x-axis. But, $\tan\theta$ also equals $\Delta y/\Delta x$ for the tangent line AB, and $\Delta y/\Delta x$ is the slope of the line. Thus, the slope of a curve at any point equals the slope of the line drawn tangent to the curve at that point. This slope, in turn, equals the derivative of y with respect to x, dy/dx, evaluated at the same point.

These applications suggest that a derivative can be visualized as the slope of a graphical plot. A derivative represents the rate of change of one quantity with respect to another. When the relationship between these two quantities is presented in graphical form, this rate of change equals the slope of the resulting plot.

The mathematics of dynamic systems involves many different operations with the derivatives of functions. In practice, derivatives of functions are not determined by plotting the functions and finding the slopes of tangent lines. Although this approach could be used, techniques have been developed that permit derivatives of functions to be determined directly based on the form of the functions. For example, the derivative of the function $f(x) = c$, where c is a constant, is zero. The graph of a constant function is a horizontal line, and the slope of a horizontal line is zero.

$$f(x) = c$$

$$\frac{d[f(x)]}{dx} = 0 \qquad (5\text{-}6)$$

The derivative of the function $f(x) = ax + c$ (compare to slope m from general form of linear equation, $y = mx + b$), where a and c are constants, is a. The graph of such a function is a straight line having a slope equal to a.

$$f(x) = ax + c$$

$$\frac{d[f(x)]}{dx} = a \qquad (5\text{-}7)$$

The derivative of the function $f(x) = ax^n$, where a and n are constants, is nax^{n-1}.

$$f(x) = ax^n$$

$$\frac{d[f(x)]}{dx} = nax^{n-1} \qquad (5\text{-}8)$$

The derivative of the function $f(x) = ae^{bx}$, where a and b are constants and e is the base of natural logarithms, is abe^{bx}.

$f(x) = ae^{bx}$

$$\frac{d[f(x)]}{dx} = abe^{bx} \tag{5-9}$$

These general techniques for finding the derivatives of functions are important for those who perform detailed mathematical calculations for dynamic systems. For example, the designers of nuclear facility systems need an understanding of these techniques, because these techniques are not encountered in the day-to-day operation of a nuclear facility. As a result, the operators of these facilities should understand what derivatives are in terms of a rate of change and a slope of a graph, but they will not normally be required to find the derivatives of functions.

The notation $d[f(x)]/dx$ is the common way to indicate the derivative of a function. In some applications, the notation $f'(x)$ is used. In other applications, the so-called dot notation is used to indicate the derivative of a function with respect to time. For example, the derivative of the amount of heat transferred, Q, with respect to time, dQ/dt, is often written as $\dot{Q}$.

It is also of interest to note that many detailed calculations for dynamic systems involve not only one derivative of a function, but several successive derivatives. The second derivative of a function is the derivative of its derivative; the third derivative is the derivative of the second derivative, and so on. For example, velocity is the first derivative of distance traveled with respect to time, $v = ds/dt$; acceleration is the derivative of velocity with respect to time, $a = dv/dt$. Thus, acceleration is the second derivative of distance traveled with respect to time. This is written as d^2s/dt^2. The notation $d^2[f(x)]/dx^2$ is the common way to indicate the second derivative of a function. In some applications, the notation $f''(x)$ is used. The notation for third, fourth, and higher order derivatives follows this same format. Dot notation can also be used for higher order derivatives with respect to time. Two dots indicates the second derivative, three dots the third derivative, and so on.

Application of Derivatives to Physical Systems

There are many different problems involving dynamic physical systems that are most readily solved using derivatives. One of the best illustrations of the application of derivatives are problems involving related rates of change. When two quantities are related by some known physical relationship, their rates of change with respect to a third quantity are also related. For example, the area of a circle is related to its radius by the formula $A = \pi r^2$. If for some reason the size of a circle is changing in time, the rate of change of its area, with respect to time for example, is also related to the rate of change of its radius with respect to time. Although these applications involve finding the derivative of function, they illustrate why derivatives are needed to solve certain problems involving dynamic systems.

Example 1:

A stone is dropped into a quiet lake, and waves move in circles outward from the location of the splash at a constant velocity of 0.5 feet per second. Determine the rate at which the area of the circle is increasing when the radius is 4 feet.

Solution:

Using the formula for the area of a circle,

$$A = \pi r^2$$

take the derivative of both sides of this equation with respect to time *t*.

$$\frac{dA}{dt} = 2\pi r \frac{dr}{dt}$$

But, *dr/dt* is the velocity of the circle moving outward which equals 0.5 ft/s and *dA/dt* is the rate at which the area is increasing, which is the quantity to be determined. Set *r* equal to 4 feet, substitute the known values into the equation, and solve for *dA/dt*.

$$\frac{dA}{dt} = 2\pi r \frac{dr}{dt}$$

$$\frac{dA}{dt} = (2)(3.1416)(4 \text{ ft})(0.5 \text{ ft/s})$$

$$\frac{dA}{dt} = 12.6 \text{ ft}^2\text{/s}$$

Thus, at a radius of 4 feet, the area is increasing at a rate of 12.6 square feet per second.

Example 2:

A ladder 26 feet long is leaning against a wall. The ladder starts to move such that the bottom end moves away from the wall at a constant velocity of 2 feet per second. What is the downward velocity of the top end of the ladder when the bottom end is 10 feet from the wall?

Solution:

Start with the Pythagorean Theorem for a right triangle: $a^2 = c^2 - b^2$

Take the derivative of both sides of this equation with respect to time *t*. The *c*, representing the length of the ladder, is a constant.

$$2a\frac{da}{dt} = -2b\frac{db}{dt}$$

$$a\frac{da}{dt} = -b\frac{db}{dt}$$

But, *db/dt* is the velocity at which the bottom end of the ladder is moving away from the wall, equal to 2 ft/s, and *da/dt* is the downward velocity of the top end of the ladder along the wall, which is the quantity to be determined. Set *b* equal to 10 feet, substitute the known values into the equation, and solve for *a*.

$$a^2 = c^2 - b^2$$

$$a = \sqrt{c^2 - b^2}$$

$$a = \sqrt{(26\ \text{ft})^2 - (10\ \text{ft})^2}$$

$$a = \sqrt{676\ \text{ft}^2 - 100\ \text{ft}^2}$$

$$a = \sqrt{576\ \text{ft}^2}$$

$$a = 24\ \text{ft}$$

$$a\frac{da}{dt} = -b\frac{db}{dt}$$

$$\frac{da}{dt} = -\frac{b}{a}\frac{db}{dt}$$

$$\frac{da}{dt} = -\frac{10\ \text{ft}}{24\ \text{ft}}(2\ \text{ft/s})$$

$$\frac{da}{dt} = -0.833\ \text{ft/s}$$

Thus, when the bottom of the ladder is 10 feet from the wall and moving at 2ft/sec., the top of the ladder is moving downward at 0.833 ft/s. (The negative sign indicates the downward direction.)

Integrals and Summations in Physical Systems

Differentials and derivatives arose in physical systems when small changes in one quantity were considered. For example, the relationship between position and time for a moving object led to the definition of the instantaneous velocity, as the derivative of the distance traveled with respect to time, *ds/dt*. In many physical systems, rates of change are measured directly. Solving problems, when this is the case, involves another aspect of the mathematics of dynamic systems; namely integral and summations.

Figure 7 is a graph of the instantaneous velocity of an object as a function of elapsed time. This is the type of graph which could be generated if the reading of the speedometer of a car were recorded as a function of time.

At any given instant of time, the velocity of the object can be determined by referring to Figure 7. However, if the distance traveled in a certain interval of time is to be determined, some new techniques must be used. Consider the velocity versus time curve of Figure 7. Let's consider the velocity changes between times t_A and t_B. The first approach is to divide the time interval into three short intervals ($\Delta t_1, \Delta t_2, \Delta t_3$), and to assume that the velocity is constant during each of these intervals. During time interval Δt_1, the velocity is assumed constant at an average velocity v_1; during the interval Δt_2, the velocity is assumed constant at an average velocity v_2; during time interval Δt_3, the velocity is assumed constant at an average velocity v_3. Then the total distance traveled is approximately the sum of the products of the velocity and the elapsed time over each of the three intervals. Equation 5-10 approximates the distance traveled during the time interval from t_a to t_b and represents the approximate area under the velocity curve during this same time interval.

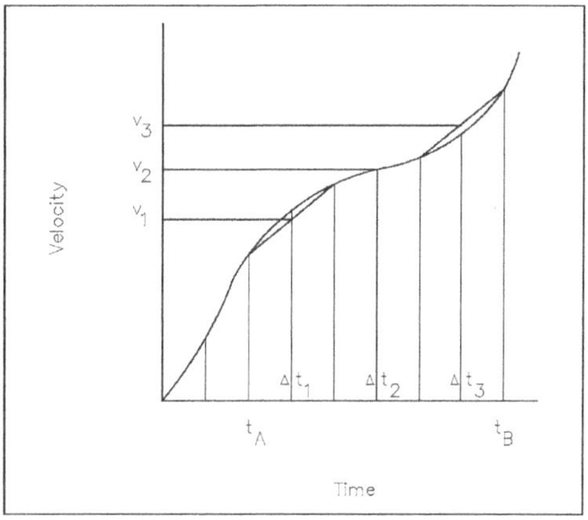

Figure 7 Graph of Velocity vs. Time

$$s = v_1\Delta t_1 + v_2\Delta t_2 + v_3\Delta t_3 \tag{5-10}$$

This type of expression is called a summation. A summation indicates the sum of a series of similar quantities. The upper case Greek letter Sigma, Σ, is used to indicate a summation. Generalized subscripts are used to simplify writing summations. For example, the summation given in Equation 5-10 would be written in the following manner:

$$S = \sum_{i=1}^{3} v_i \Delta t_i \qquad (5\text{-}11)$$

The number below the summation sign indicates the value of i in the first term of the summation; the number above the summation sign indicates the value of i in the last term of the summation.

The summation that results from dividing the time interval into three smaller intervals, as shown in Figure 7, only approximates the distance traveled. However, if the time interval is divided into incremental intervals, an exact answer can be obtained. When this is done, the distance traveled would be written as a summation with an indefinite number of terms.

$$S = \sum_{i=1}^{\infty} v_i \Delta t_i \qquad (5\text{-}12)$$

This expression defines an integral. The symbol for an integral is an elongated "s" $\int$. Using an integral, Equation 5-12 would be written in the following manner:

$$S = \int_{t_A}^{t_B} v \, dt \qquad (5\text{-}13)$$

This expression is read as S equals the integral of $v\ dt$ from $t = t_A$ to $t = t_B$. The numbers below and above the integral sign are the limits of the integral. The limits of an integral indicate the values at which the summation process, indicated by the integral, begins and ends.

As with differentials and derivatives, one of the most important parts of understanding integrals is having a physical interpretation of their meaning. For example, when a relationship is written as an integral, the physical meaning, in terms of a summation, should be readily understood. In the previous example, the distance traveled between t_A and t_B was approximated by equation 5-10. Equation 5-13 represents the exact distance traveled and also represents the exact area under the curve on figure 7 between t_A and t_B.

Examples:

1. Give the physical interpretation of the following equation relating the work, W, done when a force, F, moves a body from position x_1 to x_2.

$$W = \int_{x_1}^{x_2} F\,dx$$

The physical meaning of this equation can be stated in terms of a summation. The total amount of work done equals the integral of $F\,dx$ from $x = x_1$ to $x = x_2$. This can be visualized as taking the product of the instantaneous force, F, and the incremental change in position dx at each point between x_1 and x_2, and summing all of these products.

2. Give the physical interpretation of the following equation relating the amount of radioactive material present as a function of the elapsed time, t, and the decay constant, λ.

$$\int_{N_0}^{N_1} \frac{dN}{N} = -\lambda t$$

The physical meaning of this equation can be stated in terms of a summation. The negative of the product of the decay constant, λ, and the elapsed time, t, equals the integral of dN/N from $N = N_0$ to $n = n_1$. This integral can be visualized as taking the quotient of the incremental change in N, divided by the value of N at each point between N_0 and N_1, and summing all of these quotients.

Graphical Understanding of Integral

As with derivatives, when a functional relationship is presented in graphical form, an important understanding of the meaning of integral can be developed.

Figure 8 is a plot of the instantaneous velocity, v, of an object as a function of elapsed time, t. The functional relationship shown is given by the following equation:

$$v = 6t \qquad (5\text{-}14)$$

The distance traveled, s, between times t_A and t_B equals the integral of the velocity, v, with respect to time between the limits t_A and t_B.

$$s = \int_{t_A}^{t_B} v\,dt \qquad (5\text{-}15)$$

The value of this integral can be determined for the case plotted in Figure 8 by noting that the velocity is increasing linearly. Thus, the average velocity for the time interval between t_A and t_B is the arithmetic average of the velocity at t_A and the velocity at t_B. At time t_A, $v = 6t_A$; at time t_B, $v = 6t_B$. Thus, the average velocity for the time interval between t_A and t_B is $\frac{6t_A + 6t_B}{2}$ which equals $3(t_A + t_B)$. Using this average velocity, the total distance traveled in the time interval between t_A and t_B is the product of the elapsed time $t_B - t_A$ and the average velocity $3(t_A + t_B)$.

$$s = v_{av}\Delta t$$

$$s = 3(t_A + t_B)(t_B - t_A) \qquad (5\text{-}16)$$

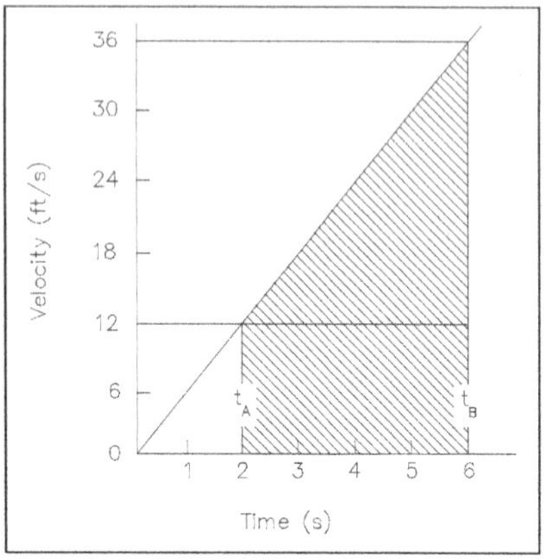

Figure 8 Graph of Velocity vs. Time

Equation 5-16 is also the value of the integral of the velocity, v, with respect to time, t, between the limits t_A -t_B for the case plotted in Figure 8.

$$\int_{t_A}^{t_B} v\,dt = 3(t_A + t_B)(t_B - t_A)$$

The cross-hatched area in Figure 8 is the area under the velocity curve between $t = t_A$ and $t = t_B$. The value of this area can be computed by adding the area of the rectangle whose sides are $t_B - t_A$ and the velocity at t_A, which equals $6t_A - t_B$, and the area of the triangle whose base is $t_B - t_A$ and whose height is the difference between the velocity at t_B and the velocity at t_A, which equals $6t_B - t_A$.

$$\text{Area} = [(t_B - t_A)(6t_A)] + \left[\frac{1}{2}(t_B - t_A)(6t_b - 6t_A)\right]$$

$$\text{Area} = 6t_A t_B - 6t_A^2 + 3t_B^2 - 6t_A t_B + 3t_A^2$$

$$\text{Area} = 3t_B^2 - 3t_A^2$$

$$\text{Area} = 3(t_B + t_A)(t_B - t_A)$$

This is exactly equal to the value of the integral of the velocity with respect to time between the limits t_A and t_B. Since the distance traveled equals the integral of the velocity with respect to time, $\int v dt$, and since this integral equals the area under the curve of velocity versus time, the distance traveled can be visualized as the area under the curve of velocity versus time.

For the case shown in Figure 8, the velocity is increasing at a constant rate. When the plot of a function is not a straight line, the area under the curve is more difficult to determine. However, it can be shown that the integral of a function equals the area between the x-axis and the graphical plot of the function.

$$\int_{X_1}^{X_2} f(x)dx = \text{Area between } f(x) \text{ and x-axis from } x_1 \text{ to } x_2$$

The mathematics of dynamic systems involves many different operations with the integral of functions. As with derivatives, in practice, the integral of functions are not determined by plotting the functions and measuring the area under the curves. Although this approach could be used, techniques have been developed which permit integral of functions to be determined directly based on the form of the functions. Actually, the technique for taking an integral is the reverse of taking a derivative. For example, the derivative of the function $f(x) = ax + c$, where a and c are constants, is a. The integral of the function $f(x) = a$, where a is a constant, is $ax + c$, where a and c are constants.

$$f(x) = a$$

$$\int f(x)dx = ax + c \qquad (5\text{-}17)$$

The integral of the function $f(x) = ax^n$, where a and n are constants, is $\frac{a}{n+1} x^{n+1} + c$, where c is another constant.

$$f(x) = ax^n$$

$$\int f(x)dx = \frac{a}{n+1} x^{n+1} + c \qquad (5\text{-}18)$$

The integral of the function $f(x) = ae^{bx}$, where a and b are constants and e is the base of natural logarithms, is $\frac{ae^{bx}}{b} + c$, where c is another constant.

$$f(x) = ae^{bx}$$

$$\int f(x)dx = \frac{a}{b}e^{bx} + c \qquad (5\text{-}19)$$

As with the techniques for finding the derivatives of functions, these general techniques for finding the integral of functions are primarily important only to those who perform detailed mathematical calculations for dynamic systems. These techniques are not encountered in the day-to-day operation of a nuclear facility. However, it is worthwhile to understand that taking an integral is the reverse of taking a derivative. It is important to understand what integral and derivatives are in terms of summations and areas under graphical plot, rates of change, and slopes of graphical plots.

Summary

The important information covered in this chapter is summarized below.

Derivatives and Differentials Summary

- The derivative of a function is defined as the rate of change of one quantity with respect to another, which is the slope of the function.
- The integral of a function is defined as the area under the curve.

end of text.

CONCLUDING MATERIAL

Review activities:

DOE - ANL-W, BNL, EG&G Idaho,
EG&G Mound, EG&G Rocky Flats,
LLNL, LANL, MMES, ORAU, REECo,
WHC, WINCO, WEMCO, and WSRC.

Preparing activity:

DOE - NE-73
Project Number 6910-0020/2

Made in the USA
Monee, IL
04 November 2026

67379248R00177